Cambridge
checkpoint

Lower Secondary
Mathematics
WORKBOOK
9

Ric Pimentel
Frankie Pimentel
Terry Wall

HODDER EDUCATION

Contents

Section 3

Introduction

Welcome to *Cambridge Checkpoint Lower Secondary Mathematics Workbook Stage 9*. This is the third of three books intended to provide practice in the skills you have acquired by using the Cambridge Checkpoint Lower Secondary Mathematics series of student's books. Each of the workbooks is planned to complement the corresponding student's book, which is split into three sections containing units from the four areas of the Cambridge Lower Secondary Mathematics curriculum framework. This workbook is intended to help you acquire the skills to be fully competent in your mathematics. Just as the *Student's Book* included green, amber and red questions in the exercises to indicate the level of the question, this workbook includes a traffic light system to highlight the same levelling.

This symbol indicates the green level from the *Student's Book*. They are introductory questions.

This symbol indicates the amber level. They are more challenging questions.

This symbol indicates the red level. They are questions to really challenge yourself.

This star shows that you will be thinking and working mathematically (TWM).

There is no set way to approach using this workbook – you may wish to use it to supplement your understanding as you work through the *Student's Book*, or you may prefer to use it to recap on particular topics. It is hoped that the organisation of the material in the book is flexible enough for whichever approach you prefer.

So now it is time to start. Read each question carefully, think about it, then write down the answer in the space provided or as instructed. Ask for help if you need it but try hard first. Try to learn by thinking and working mathematically. Write down what you are thinking so that others can understand what you have done and help to correct your mistakes.

SECTION 1

1 Indices and standard form

Exercise 1.1

1 Write each of the following as a fraction.

a 4^{-2} = ...

b 5^{-3} = ...

c 2^{-4} = ...

2 Write each of these as an integer or a fraction.

a 5×5^{-1} = ...

b 8×2^{-3} = ...

c 400×2^{-4} = ...

d 64×4^{-5} = ...

e $3^5 \times 3^{-3}$ = ...

f $4^3 \times 4^{-6} \times 4$ = ...

3 Using indices, find the value of each of these. Give your answers in positive index form.

a $6^3 \times 6^{-5}$ = ...

b $9^{-5} \times 9^2$ = ...

c $4^3 \div 4^6$ = ...

4 Match each fraction with its corresponding number written in index form. For those that do not have a pair, write the correct answer in the empty box to match it up.

4^{-2}		7^{-2}	6^{-3}	2^{-6}

$\frac{1}{216}$	$\frac{1}{16}$	$\frac{1}{49}$		$\frac{1}{27}$

5 A cube has a side length of $\frac{1}{5}$ cm.

a Give a **convincing** explanation to show that its volume is 5^{-3} cm^3.

....

....

....

b Show that its total surface area is $\frac{6}{5^2}$ cm^2.

....

....

....

6 Find the volume of the cuboid below, giving your answer in the form 2^a, where a is an integer.

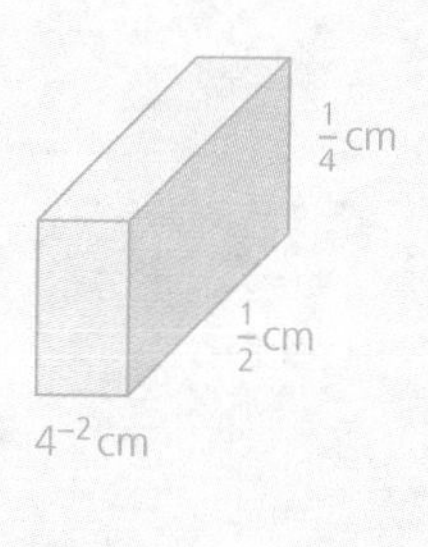

....

....

....

Exercises 1.2–1.3

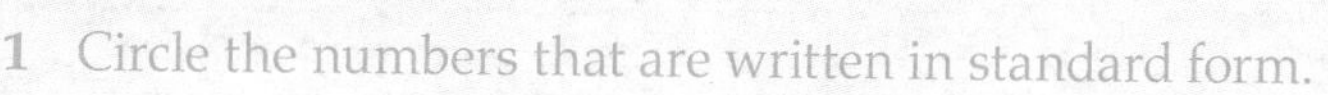

1 Circle the numbers that are written in standard form.

5.7×10^3 $\quad$ 32.6×10^5 $\quad$ $2.5 \div 10^8$ $\quad$ 0.7×10^4 $\quad$ 4.82×10^{-5}

2 Write the following numbers in standard form.

a 3200 = ...

b 5 040 000 = ...

c 730 = ...

d 0.000 349 = ...

e 0.000 020 7 = ...

f 0.444 = ...

3 The following table shows the populations of three cities in India. Which city has the greatest population? Justify your answer.

City	Population
Jodhpur	1.1 million
Chennai	8.7×10^6
Mangalore	550000

4 Using a calculator, work out the following. Give your answers in standard form, correct to 2 significant figures.

a $(6.3 \times 10^3) + (5.7 \times 10^5)$

b $(4.8 \times 10^7) - (9.3 \times 10^1)$

c $(7.2 \times 10^{-2}) \times (1.5 \times 10^{-4})$

5 The mass of an electron is 9.1×10^{-28} grams. What is the total mass of 4×10^8 electrons?

2 Pythagoras' theorem

Exercises 2.1–2.3

1 Use Pythagoras' theorem to calculate the length of the hypotenuse in each of these diagrams. Give your answers correct to 1 decimal place.

a

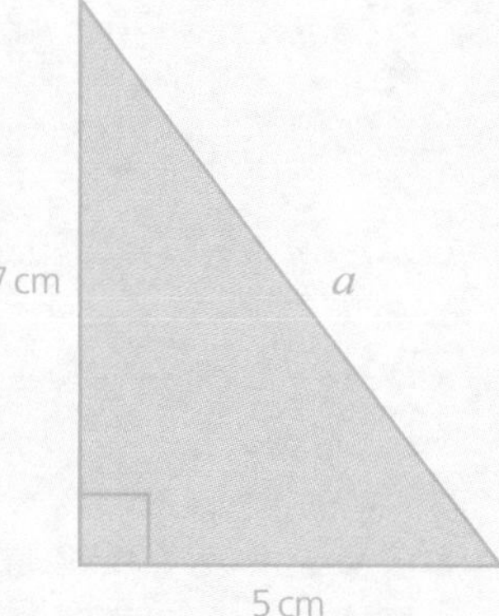

....

....

....

b

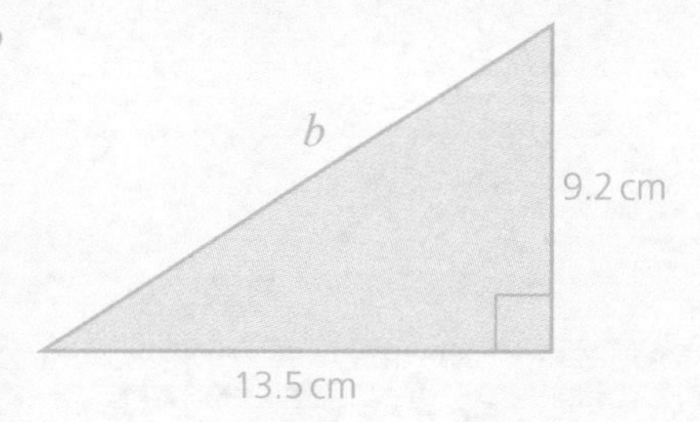

....

....

....

2 Use Pythagoras' theorem to calculate the length of the unknown side in each of these diagrams. Give your answers correct to 1 decimal place.

a

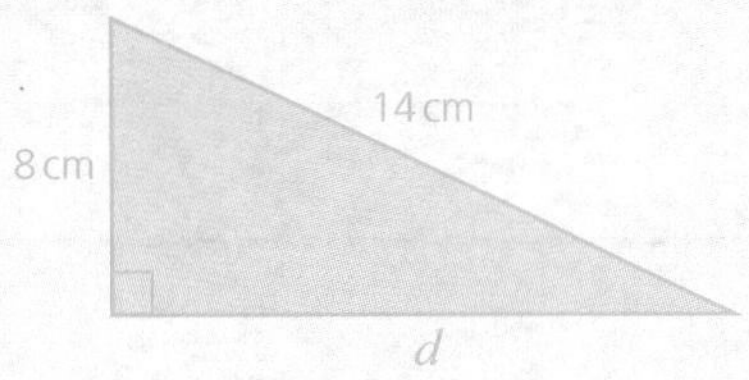

8 cm

14 cm

d

....

....

....

b

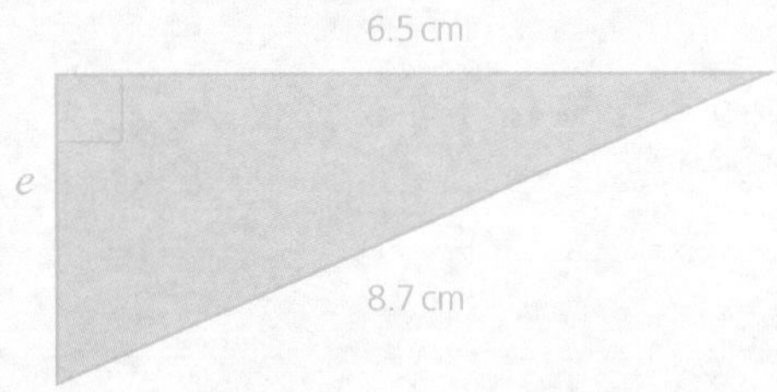

6.5 cm

e

8.7 cm

....

....

....

c

f

0.5 cm

0.8 cm

....

....

....

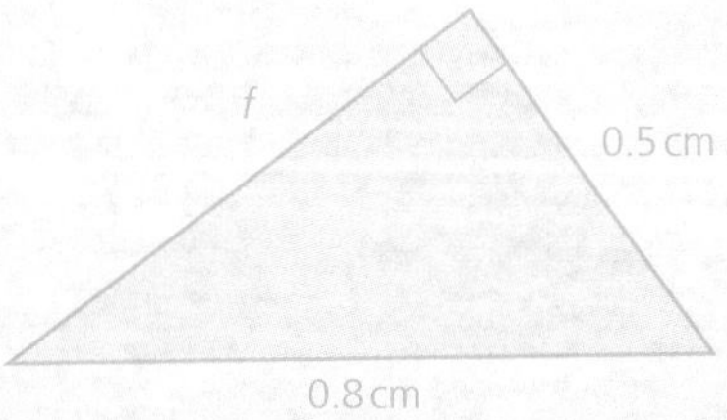

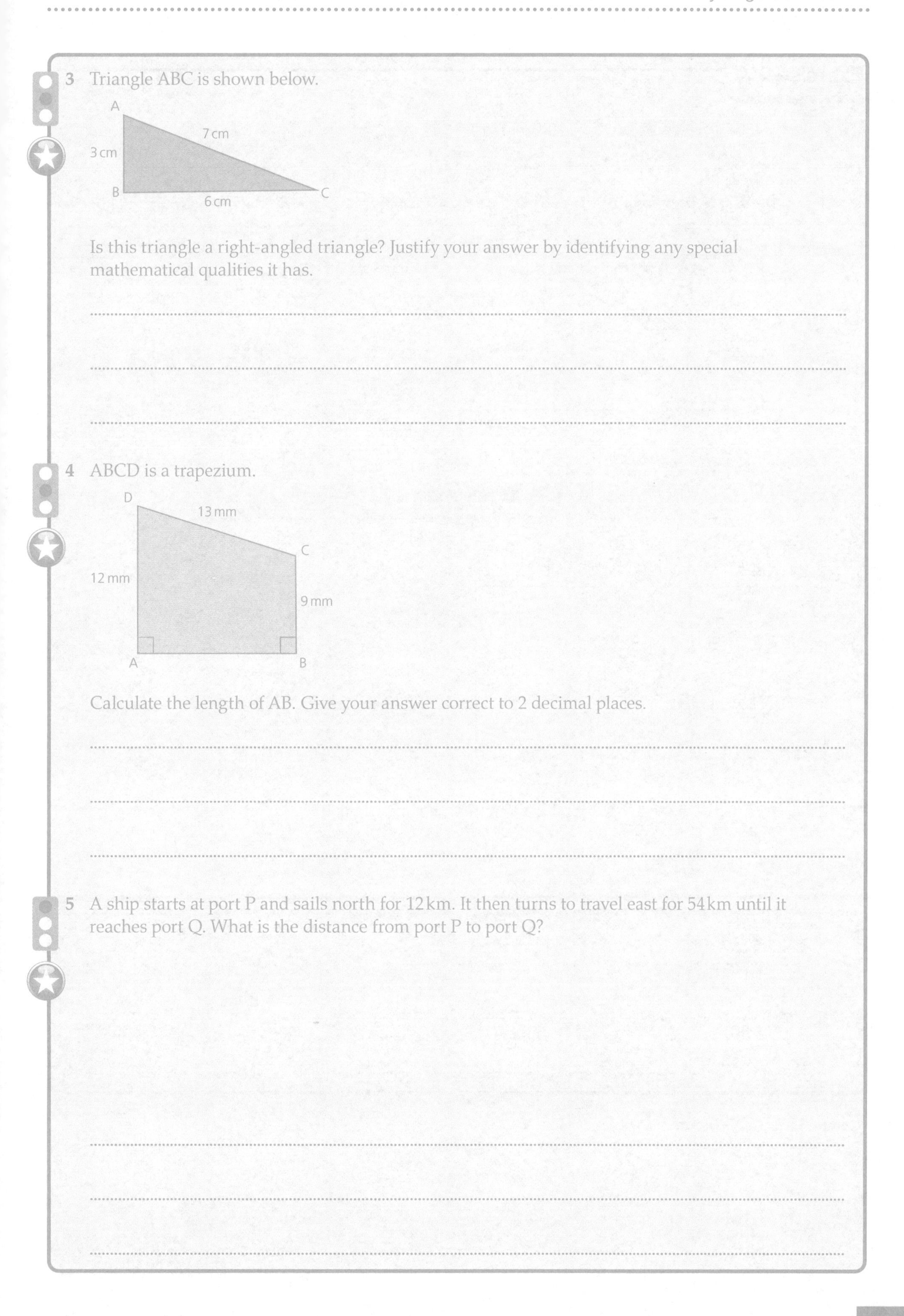

3 Triangle ABC is shown below.

Is this triangle a right-angled triangle? Justify your answer by identifying any special mathematical qualities it has.

....

....

....

4 ABCD is a trapezium.

Calculate the length of AB. Give your answer correct to 2 decimal places.

....

....

....

5 A ship starts at port P and sails north for 12 km. It then turns to travel east for 54 km until it reaches port Q. What is the distance from port P to port Q?

....

....

....

Exercise 2.4

1 Find the lengths of sides r and t in this diagram.

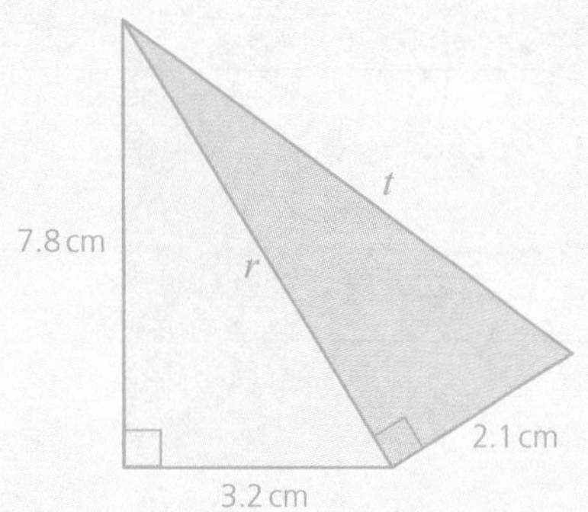

2 A rectangular envelope is shown below.

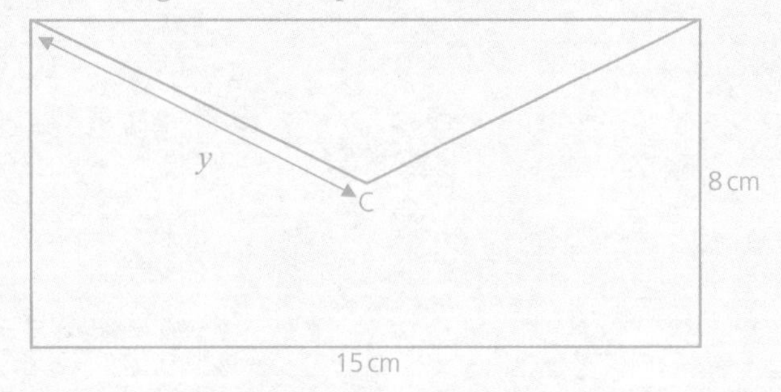

C is the centre of the rectangle. Find the length of the line labelled y.

3 Data collection and sampling

Exercise 3.1

The questions below are about collecting data to answer this question:

'Are people buying new mobile phones too often?'

1 **Improve** the question to make it more precise.

....

....

2 What primary data could be collected to answer the question?

....

3 Give an example of secondary data that could be collected.

....

4 Suggest three questions that could be used in a questionnaire.

1 ....

2 ....

3 ....

5 What sample size would you use for your questionnaire? Give a reason for your answer.

....

....

6 How would you select your sample?

....

....

7 What should be done prior to distributing a questionnaire? Give a reason for your answer.

....

....

4 Area and circumference of a circle

Exercise 4.1

1 Calculate the circumference of each of these circles. Give your answers correct to 2 decimal places.

a

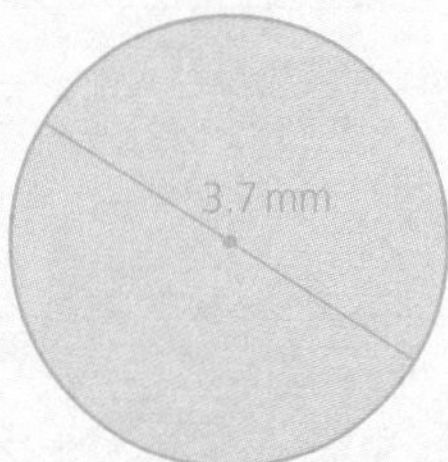

....

....

b

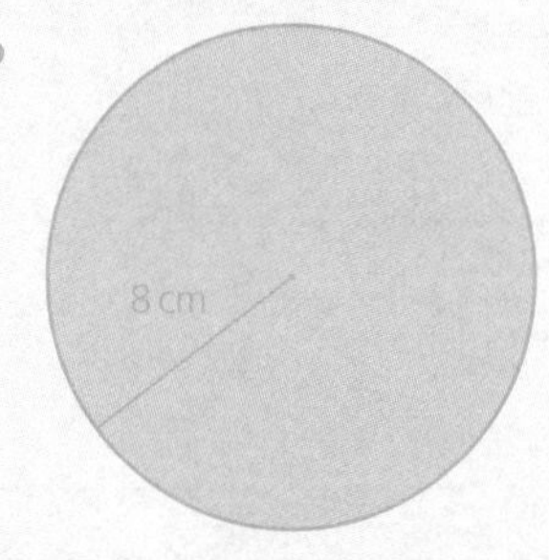

....

....

2 Calculate the perimeter of this shape.

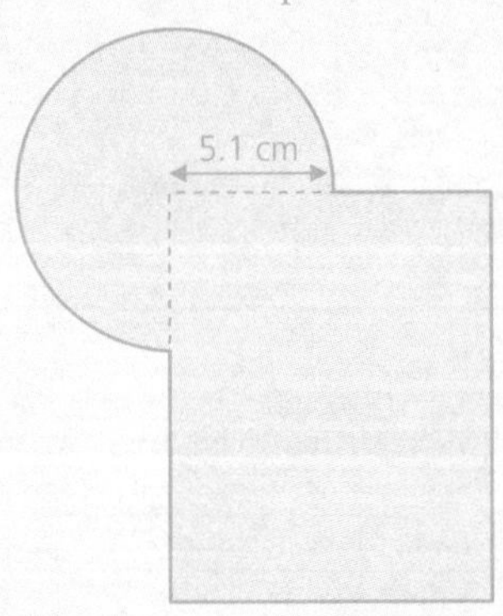

....

....

....

....

3 The wheels on Rimsha's bike have a diameter of 48 cm. She travels so that the wheels go round completely 15 times. How far does Rimsha travel? Give your answers to 1 decimal place.

....

....

....

....

Exercises 4.2–4.3

1 Calculate the area of each of these circles. Give your answers correct to 1 decimal place.

a

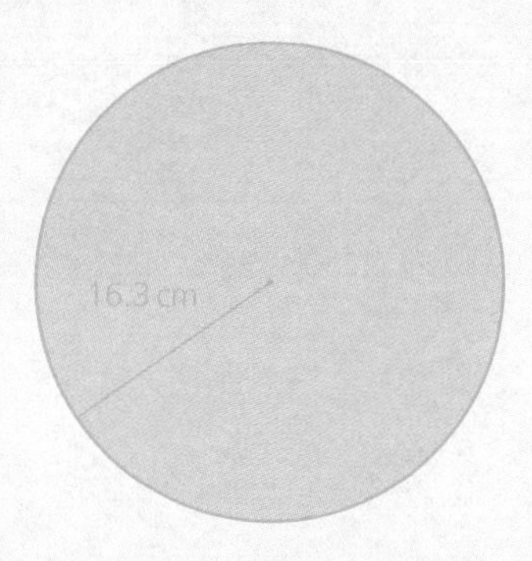

....

....

b

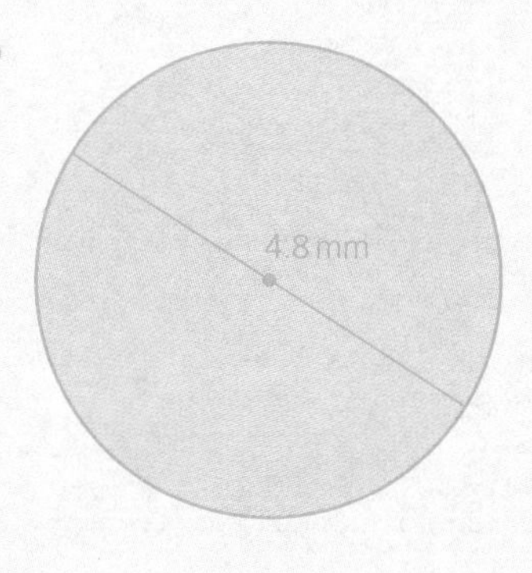

....

....

2 Calculate the area of this shape. Give your answer correct to 1 decimal place.

....

....

....

3 This diagram shows a garden with a pond.

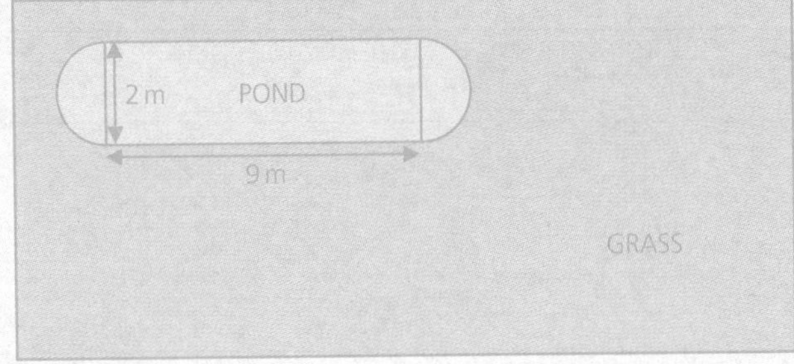

a Calculate the area of the garden which is grass.

....

....

....

....

b A gardener wants to put a fence around the whole garden and around the pond. What length of fencing will he need? Give your answer correct to the nearest metre.

....

....

....

....

c A metre of fencing costs \$17.50. Using your answer to part (b), how much will it cost the gardener to put a fence around the whole garden and around the pond?

....

....

5 Order of operations with algebra

Exercise 5.1

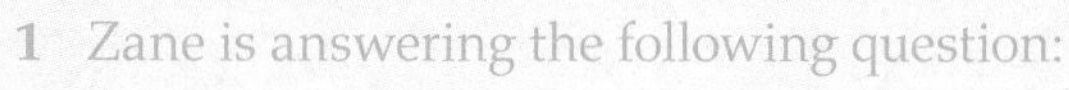

1 Zane is answering the following question:

Evaluate $\frac{(4+2)^2}{3}+3\times5$

Here are his steps to working it out:

i $\frac{(6)^2}{3}+3\times5$ **ii** $\frac{12}{3}+3\times5$ **iii** $4+3\times5$ **iv** 7×5 **v** 35

a Zane's teacher says he has made two mistakes. What are they? Justify your answers.

1 ....

....

2 ....

....

b Calculate the correct answer.

....

....

2 Evaluate the following.

a $\frac{8}{2}-3^2\times2=$

....

....

b $\left(\frac{10}{2}\right)^2-4\times3=$

....

....

c $\left(\frac{16-2}{2}\right)^2+8\times5=$

....

....

3 Evaluate the following when $x = 3$ and $y = 5$.

a $y^2 + \frac{x(x+1)}{2}$

b $(2x+1)(3y-2) - \left(\frac{10x}{2}\right)$

c $2y^2 - \frac{3x+1}{2} \times x^2$

6 Large and small units

Exercise 6.1

1 Taking the speed of light as 3.0×10^8 m/s, convert the following into km.

a 2.5 light years

....

b 4.75 light years

....

c 10.325 light years

....

2 Write the following lengths in µm.

a 0.000 002 5 m = ...

b 8.3×10^{-10} m = ...

c 0.43 mm = ...

3 Write the following lengths in nm.

a 7.5 µm = ...

b 4.0×10^{-12} m = ...

c 1.3×10^{-5} mm = ...

4 Put the following lengths in ascending order.

250 µm   700 nm 7×10^{-12} m   0.000 000 005 6 m   52 µm   0.06 mm

....

....

Exercise 6.2

1 The mass of a lorry is 44 tonnes and the mass of a boat is 35 000 kg. Which is heavier? Justify your answer.

....

....

2 Calculate the following.

a How many kilograms are there in 4.5 t? ....

b How many grams are there in 0.2 t? ....

c How many milligrams are there in 30 g? ....

d How many micrograms are there in 2 mg? ....

e How many grams are there in 34 000 μg? ....

3 1000 cm^3 of water is equivalent to 1 litre; 1 litre of water has a mass of 1 kg. Calculate the mass of water that would completely fill each of these containers. Give your answers in g.

a

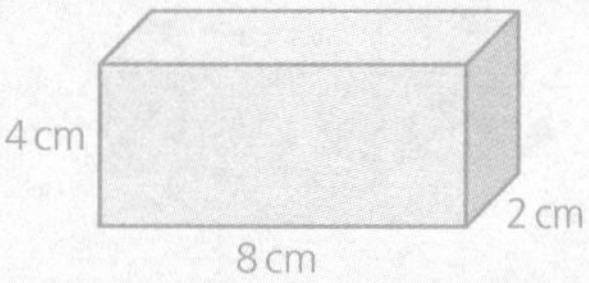

4 cm
2 cm
8 cm

....

....

b

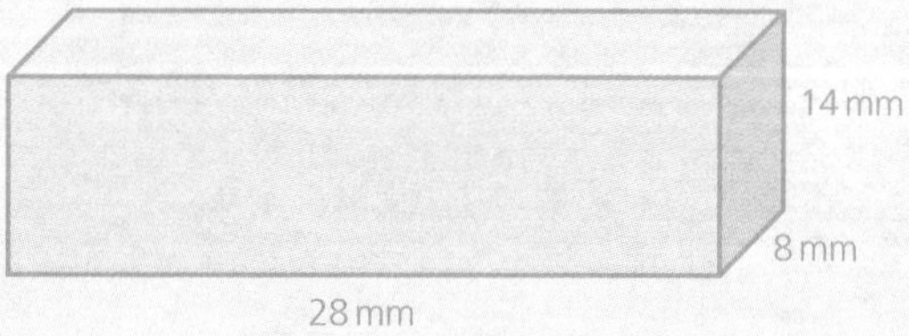

14 mm
8 mm
28 mm

....

....

4 The mass of $1\,m^3$ of water is given as 1 tonne.

The mass of 1 litre of water is 1 kg.

The diagram below shows a water tank in the shape of a triangular prism.

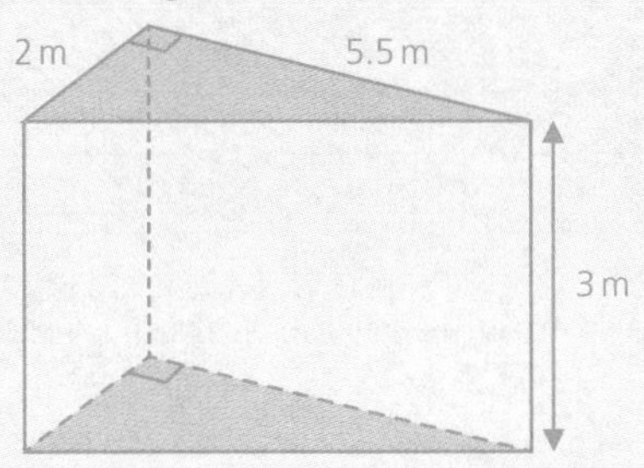

a Calculate the mass of water that could fill this tank.

....

....

b The tank is empty and fills at a rate of 250 litres per minute. How long would it take to fill the tank?

....

....

....

Exercise 6.3

1 Convert the following.

a 20 000 000 B to MB..........

b 650 MB to GB..........

c 20 GB to TB..........

2 On average a movie will take up 1.3 GB of storage space. How many movies could I store if I had 0.4 TB of space?

....

....

3 Ekon's smartphone has 16 GB of storage space. The space is currently used for the following:

- System files = 6 GB
- Messaging service = 4.2 GB
- Music streaming = 450 MB
- Social media apps = 1.3 GB
- Game apps = 800 MB

Ekon wants to download an online video-calling app that will take up 2.5 GB of storage space. He **conjectures** that he has enough space to download the app. Is he correct?

....

....

....

4 Hania is watching her favourite TV series. Each episode takes 25 minutes to download and occupies 350 MB on her laptop. Hania has decided to download a total of 11.2 GB of episodes. She starts the download at 11 p.m. on Friday night and leaves her laptop to complete the full download. When will the laptop have finished the download?

....

....

....

7 Record, organise and represent data

Exercise 7.1

1 The table below shows the heights (cm) of ten sunflowers and the total number of days of bright sunlight each sunflower has had.

Number of days of bright sunlight	35	30	28	35	33	40	38	50	45	32
Height (cm)	147	123	118	150	135	180	177	182	165	160

a What type of graph will be useful to show the relationship between the two variables?

....

b Plot a graph of height against number of days of bright sunlight using the axes below.

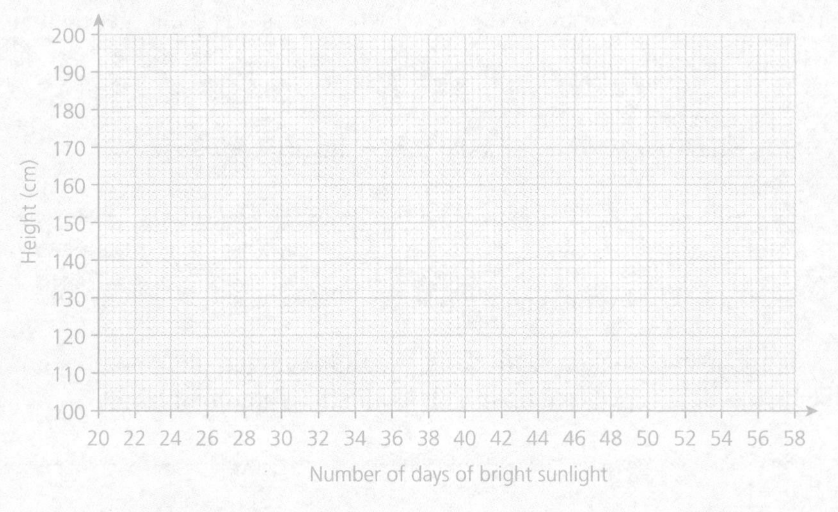

c Draw a line of best fit for the data.

d Use your graph to predict the height of a sunflower which has had 42 days of bright sunlight.

....

e Use your graph to predict how many days of bright sunlight a 135 cm tall sunflower has had.

....

f A student states this graph can be used to work out the height of any sunflower, as long as we know how many days of bright sunlight it has had. Comment on this statement.

....

....

....

2 This graph shows the distribution of the lengths of snakes of two different species, P and Q. The mean length of a snake from species P is M_1; the mean length of a snake from species Q is M_2.

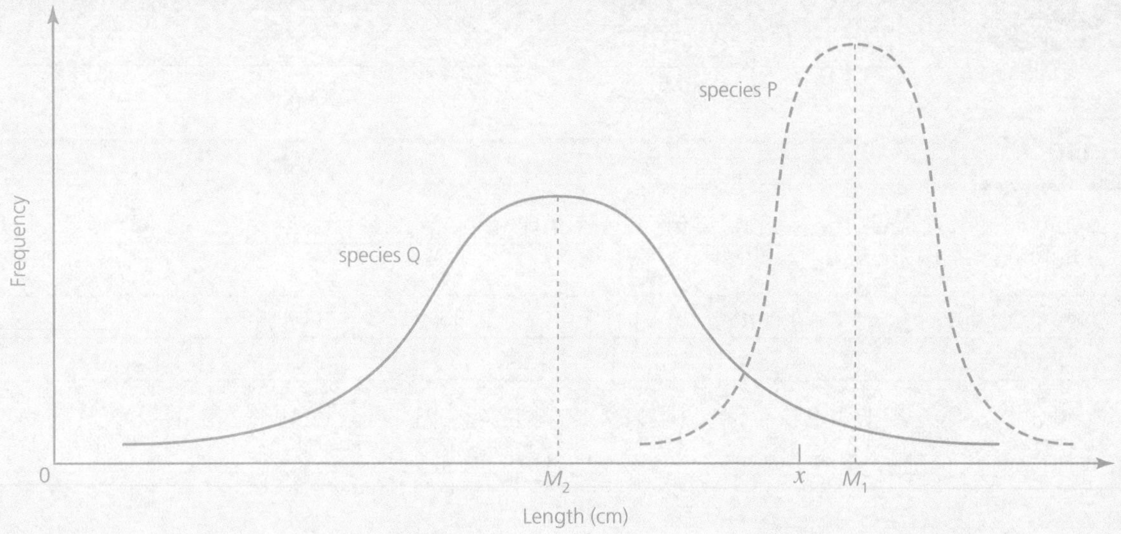

a By looking at the characteristics of both graphs, decide which species has the larger mean length. ....

b Which species has lengths which are within a smaller range? Justify your answer.

....

....

c Another snake is caught and its length is measured. Its length is x cm (as shown on the graph). Is this snake more likely to be from species P or species Q? Justify your answer.

....

....

3 The back-to-back stem-and-leaf diagram shows the heights (cm) of 20 tomato plants. Ten of the tomato plants were growing outside and the other ten tomato plants were growing in a greenhouse.

		A			B	
	9	8	1	2	3	6
8	7	7	2	1	2	2
5	4	3	3	4	5	5
	6	2	4	0		

Key: 7|2|1 = 27cm for A and 21cm for B

a Why is a back-to-back stem-and-leaf diagram useful?

....

....

b Compare the distribution of results from both groups of tomato plants, commenting on any similarities and/or differences.

....

....

....

c Which group, A or B, do you think were the plants growing in the greenhouse? Give a **convincing** reason for your answer.

....

....

....

8 Surface area and volume of prisms

Exercise 8.1

1 Calculate the volume of each of these cylinders, where r = radius of circle, d = diameter and l = length. Give your answers in cm^3.

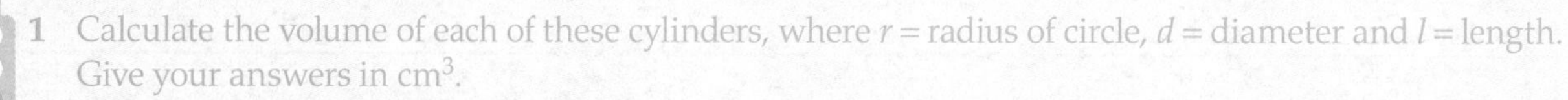

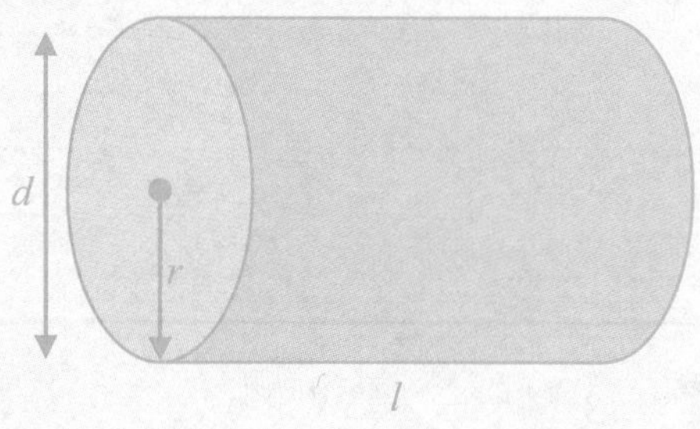

a $d = 10\,cm$ and $l = 0.6\,m$

....

....

b $r = 7.5\,mm$ and $l = 1.2\,cm$

....

....

2 Calculate the volume of the prism.

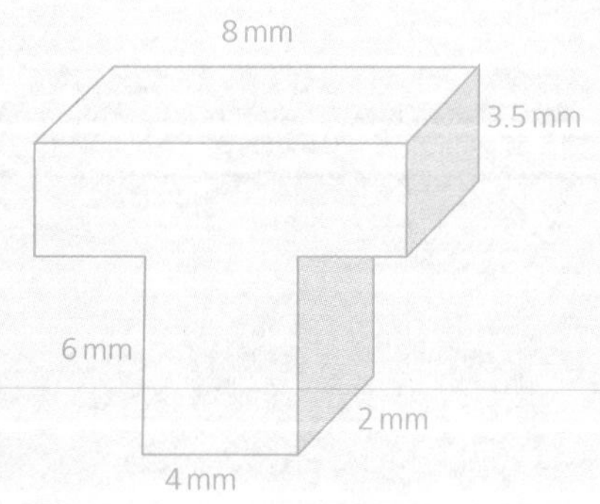

....

....

....

....

3 Part of a concrete tunnel is shown. The inner radius is 12 m, the outer radius is 15 m and the tunnel is 120 m long.

Calculate the volume of concrete used in making the tunnel. Give your answer in m^3 and correct to 3 significant figures.

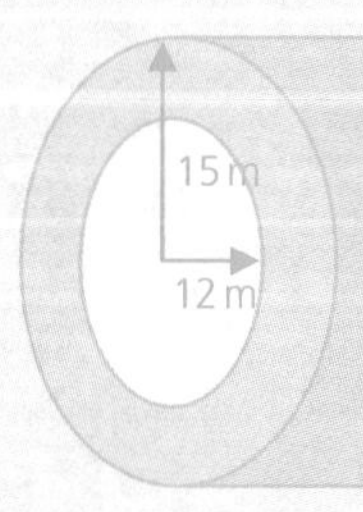

4 Here is a cylindrical water tank that has a radius of 15 cm and a height of 65 cm. The water in it has a depth of 30 cm.

15 cm
65 cm
30 cm

Asif has 200 cubes, each with a side length of 5 cm. He **conjectures** that he can put them all in the tank without the water overflowing. Is he correct? Justify your answer.

Exercise 8.2

1 Calculate the surface area of the following cylinder.

3.7 m
6.2 m

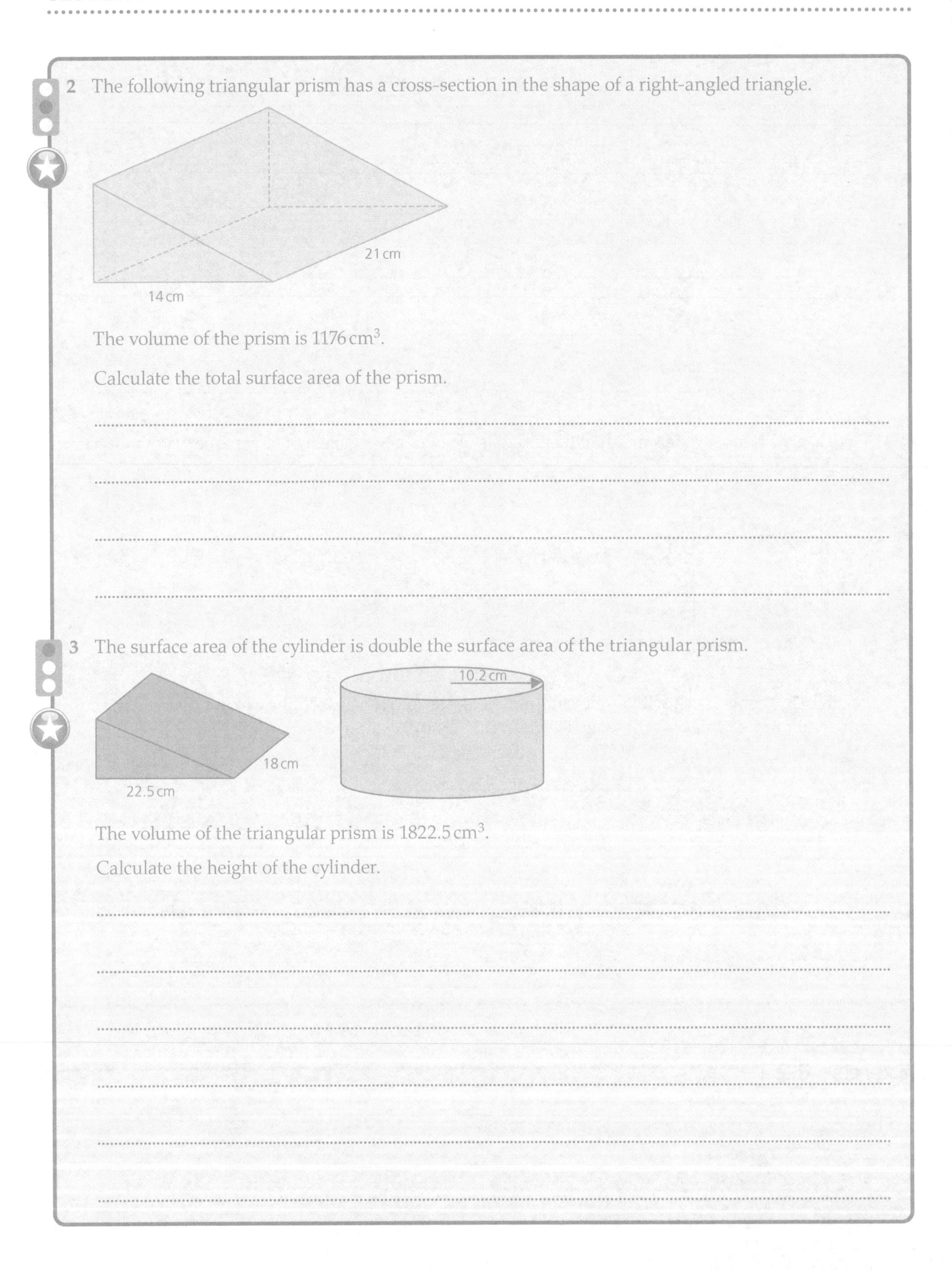

2 The following triangular prism has a cross-section in the shape of a right-angled triangle.

The volume of the prism is $1176\,\text{cm}^3$.

Calculate the total surface area of the prism.

3 The surface area of the cylinder is double the surface area of the triangular prism.

The volume of the triangular prism is $1822.5\,\text{cm}^3$.

Calculate the height of the cylinder.

9 Rational and irrational numbers

Exercise 9.1

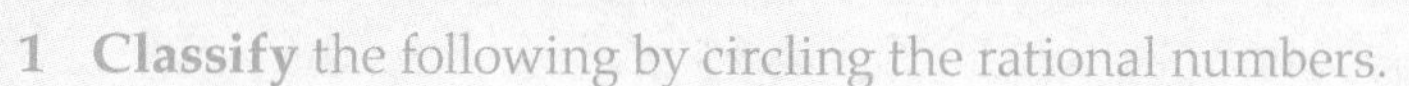

1 **Classify** the following by circling the rational numbers.

$\sqrt{12}$ $\quad$ $5.\dot{6}$ $\quad$ -3.2 $\quad$ $\sqrt{25}$ $\quad$ 3π

2 Using a calculator, calculate the following to decide whether each answer is a rational or irrational number.

a $\sqrt{8} \times \sqrt{10} =$ ....

b $\sqrt{4} \times \sqrt{16} =$ ....

c $\sqrt{200} \div \sqrt{8} =$ ....

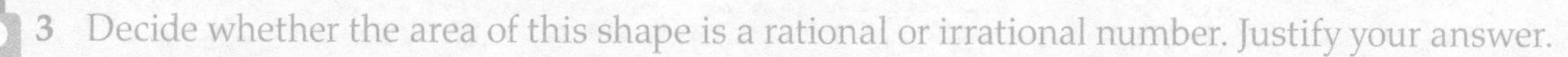

3 Decide whether the area of this shape is a rational or irrational number. Justify your answer.

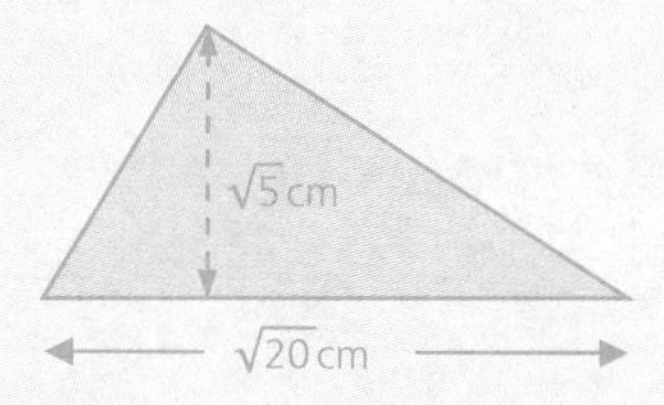

....

....

4 Prove whether or not the triangle below is right-angled.

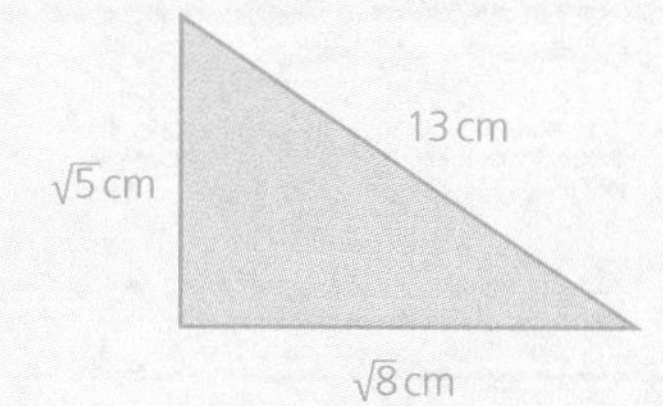

....

....

10 Mutually exclusive events

Exercise 10.1

1 There are blue, red and green marbles in a bag. The probability of getting each of the three colours of marbles is as follows:

Colour	Blue	Red	Green
Probability	0.3	0.5	0.1

Explain why the probability table must be incorrect.

....

....

....

2 A spinner is split into five different-sized sectors, labelled 1–5.

It is known that when the spinner is spun then:

$P(1) = 0.15$
$P(2) = 0.32$
$P(3) = 0.08$

It is known that the probability of the spinner landing on 5 is twice as likely as it landing on 4.

a Calculate P(4). ....

b Calculate P(5). ....

c If the spinner was spun 300 times, how many times would it be expected to land on the number 2?

....

....

3 A marshmallow company produces marshmallows in five different flavours: vanilla, strawberry, coconut, apple and rose.

One day they take 500 marshmallows to sell at a summer fair; 180 of them are vanilla flavoured.

On the same day they take with them:
– 50 more strawberry-flavoured marshmallows than apple ones
– twice as many coconut-flavoured marshmallows as apple ones
– 20 fewer rose-flavoured marshmallows than strawberry ones.

Customers pick a marshmallow at random.

Calculate:

a the probability that the first marshmallow to be picked is strawberry flavoured

....

....

b the probability that the first marshmallow to be picked is not coconut flavoured.

....

....

SECTION 2

11 Rounding and estimating numbers

Exercise 11.1

1 Give the upper and lower limits for the following approximations.

a 3000 rounded to the nearest 1000

Upper limit = ...

Lower limit = ...

b 600 rounded to the nearest 10

Upper limit = ...

Lower limit = ...

c −3 rounded to the nearest 1

Upper limit = ...

Lower limit = ...

2 A newspaper reports that 1500 people attended a show at the theatre.

a The newspaper rounded the actual figure to the nearest 100 people. What is the maximum number of people that could have attended the theatre?

....

b The theatre states that they sold 1570 tickets to the show. Did the newspaper report their rounded figure correctly? Give a reason for your answer.

....

3 Each of the following numbers is correct to 1 decimal place. Using x as the number, express the range in which the number lies as an inequality.

a 4.5 ....

b 9.2 ....

c 70.0 ....

4 Each of the following numbers is correct to 2 significant figures. Using x as the number, express the range in which the number lies as an inequality.

a 3.7 ....

b 720 ....

c 0.094 ....

Exercise 11.2

1 Work out the answers to the following calculations.

a $6.2 \times 10^3 =$ ......

b $14 \div 10^2 =$ ......

c $0.72 \times 10^0 =$ ......

d $143 \times 10^{-2} =$ ......

e $0.93 \div 10^{-4} =$ ......

f $-0.03 \times 10^{-1} =$ ......

2 Tick the boxes next to the calculations that will give an answer larger than 5.

a 5×10^{-2} ☐ b 5×10^2 ☐ c $5 \div \frac{1}{10^{-2}}$ ☐

d $5 \times \frac{1}{10^3}$ ☐ e $5 \div 10^3$ ☐ f $5 \div \frac{1}{10^2}$ ☐

3 A bookshelf is 1.2 m long. Each book is 3.2 cm thick.

a Write down a calculation to work out how many books would fit on the shelf.

......

b Write down another calculation, using powers of 10, that would give an equivalent answer.

......

c How many books would fit on the shelf?

......

4 Find the value of x for the following calculations.

a 6.3×10^x is one hundred times bigger than 6.3×10^2.

...... $x =$ ......

b 0.7×10^x is ten times bigger than 0.7×10^{-4}.

...... $x =$ ......

c $41.6 \div 10^3$ is one thousand times bigger than $41.6 \div 10^x$.

...... $x =$ ......

Exercise 11.3

1 For the following calculations:

i work out an estimate for the answer

ii calculate the answer.

a 53.7×14

Estimate = ...

Answer = ...

b -72×3.7

Estimate = ...

Answer = ...

c $-74.52 \div 23$

Estimate = ...

Answer = ...

2 A 36.8 m length of ribbon is to be shared equally between eight people. Calculate to the nearest cm how much each person will receive.

....

....

3 Decide whether the following statement is true or false. Justify your answer.

$35.7 \times 0.32 = 3.57 \times 32$

....

....

4 A lorry can travel an average of 37.1 km per litre of diesel.
The cost of diesel is $1.48 per litre.

a Estimate the cost of travelling a distance of 500 km.

....

....

....

....

b Calculate the exact cost of travelling the 500 km journey.

....

....

....

12 Further data interpretation

Exercise 12.1

1 The heights (h) of some seedlings are recorded in the table below.

Height (cm)	$0 \leqslant h < 2$	$2 \leqslant h < 4$	$4 \leqslant h < 6$	$6 \leqslant h < 8$
Frequency	3	5	6	4

a Explain whether the height h is discrete or continuous data.

....

....

b A gardener **conjectures** that the average height of these seedlings will be greater than 4.2 cm.

i Is the gardener's prediction correct?

....

ii Justify your answer to part (i).

....

....

2 As part of their training, the runners in two relay teams record their times over a week for running 100 m. Their times for completing 100 m are shown in the table.

Time (seconds)	Relay team A	Relay team B
$9.5 < T \leqslant 9.7$	3	2
$9.7 < T \leqslant 9.9$	5	7
$9.9 < T \leqslant 10.1$	3	2
$10.1 < T \leqslant 10.3$	0	1
$10.3 < T \leqslant 10.5$	1	0

a Compare the performances of the two relay teams, by considering the mean, median, mode and range of their data.

....

....

....

....

....

....

....

b Which relay team is the better team? Give reasons for your answer.

....

....

....

....

3 Each of the following sets of grouped continuous data has one frequency value missing, labelled N.

i Decide on a possible value for N which satisfies the condition stated.

ii Justify your choice of value.

a The modal class is $4 < x \leqslant 8$.

Class interval	Frequency
$0 < x \leqslant 4$	5
$4 < x \leqslant 8$	N
$8 < x \leqslant 12$	3
$12 < x \leqslant 16$	2

i ....

ii ....

....

b The class where the median lies is $4 < x \leqslant 8$.

Class interval	Frequency
$0 < x \leqslant 4$	3
$4 < x \leqslant 8$	7
$8 < x \leqslant 12$	N
$12 < x \leqslant 16$	3

i ....

ii ....

....

c The estimated mean is 7.4.

Class interval	Frequency
$0 < x \leqslant 4$	4
$4 < x \leqslant 8$	N
$8 < x \leqslant 12$	5
$12 < x \leqslant 16$	3

i ....

ii ....

....

Exercise 12.2

1 The pie charts below compare the ages of people living in two different cities, A and B.

Population and age breakdown in City A

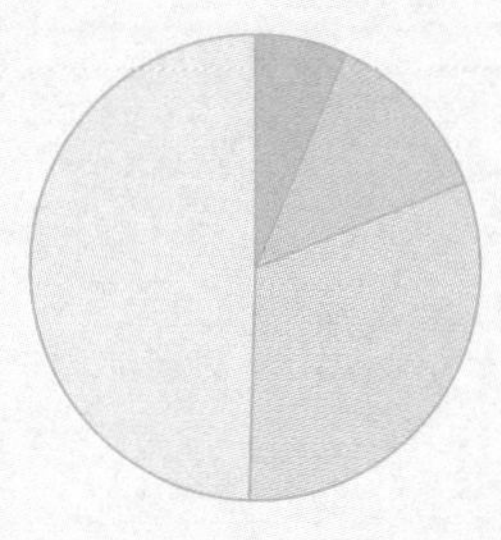

■0–19 ■20–39 ■40–59 ■60+

Population and age breakdown in City B

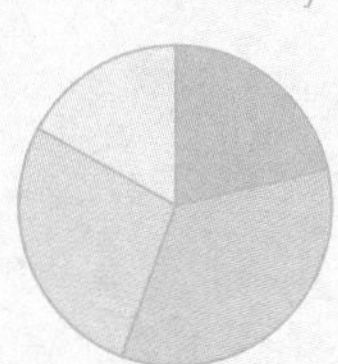

■0–19 ■20–39 ■40–59 ■60+

a What do these graphs suggest about the size of the population in these places in comparison to each other?

....

....

b As the population in City A is, in fact, twice the population of City B, the radius of the pie chart has been drawn twice as big. By **critiquing** the charts, explain why they might be misleading.

....

....

c Sofia states that the number of 20–39 year olds living in City B is higher than the number living in City A. Give a reason why she could be wrong.

....

....

13 Further transformations

Exercise 13.1

1 For each of these diagrams of three-dimensional objects, draw a plane of symmetry and work out how many planes of symmetry the shape has in total.

a A right-angled triangular prism

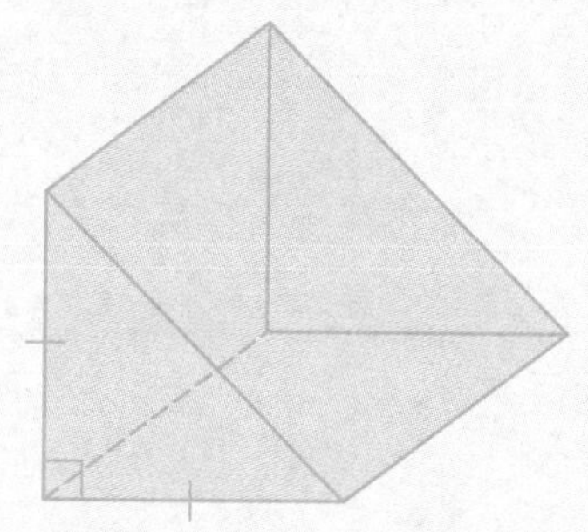

Number of planes of symmetry = ...

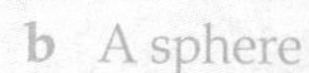

b A sphere

Number of planes of symmetry = ...

2 A cuboid is made from six cubes as shown.

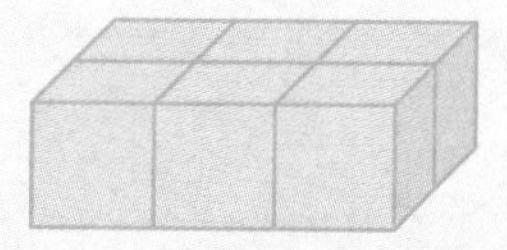

a How many planes of symmetry does the cuboid have? ....

b Draw all the planes of symmetry, each on a separate diagram.

c Four extra cubes are added to the shape. Draw one position for the four cubes so that the new shape has just two planes of symmetry. Draw the planes of symmetry on your diagram.

Exercise 13.2

1 Enlarge the following shape by a scale factor 2, centre of enlargement O. Label the vertices of the enlarged shape using the correct notation.

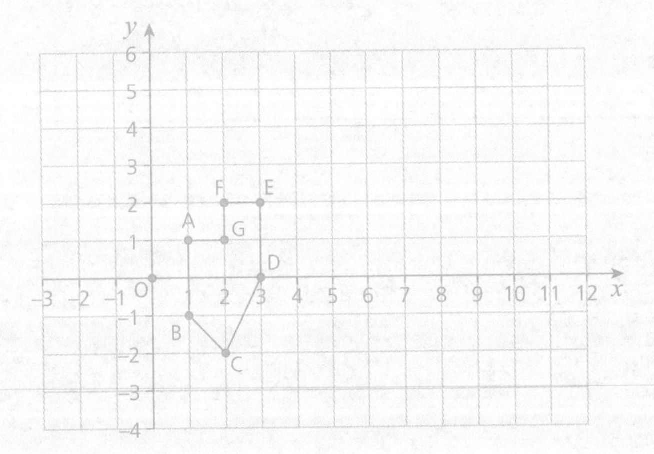

2 The coordinates of the vertices of a triangle PQR are given as P(1, –2), Q(4, –1) and R(0, 1). The shape is enlarged by a scale factor 4 from a centre of enlargement (2, 0). Calculate the coordinates of P′, Q′ and R′.

Exercises 13.3–13.4

1 The diagram shows shape X.

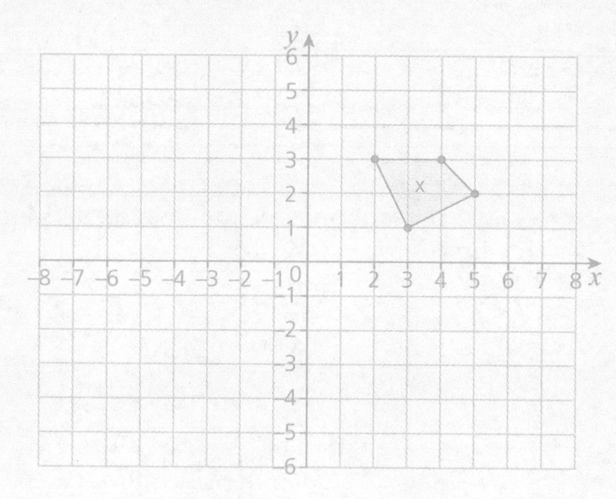

Shape X is translated $\begin{pmatrix} 2 \\ -3 \end{pmatrix}$ to give shape Y. Shape Y is reflected in the line $x = 2$ to give shape Z.

Draw shape Y and shape Z on the diagram.

2 The diagram shows shape X.

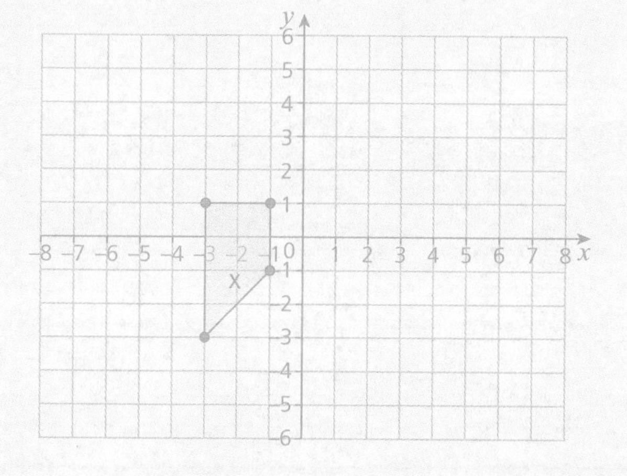

a Using two different transformations, draw a congruent shape to shape X.

b Describe your two transformations.

....

....

....

c What transformation could you not use to get a congruent shape? Justify your answer.

....

....

14 Further fractions and decimals

Exercise 14.1

1 Use a calculator to work out whether these fractions produce a terminating decimal or a recurring decimal.

a $\frac{6}{10}$ ....

b $\frac{4}{9}$ ....

c $\frac{3}{11}$ ....

Exercise 14.2

In questions 1–3 work out the answer to the calculations, showing your method clearly and giving your answers as:

a a mixed number

b an improper fraction.

1 $3\frac{1}{5} - 1\frac{1}{10} =$

....

....

....

....

2 $5\frac{7}{8} - \left(\frac{2}{5} + \frac{8}{15}\right) =$

....

....

....

....

3 $\frac{7}{5}+\left(\frac{4}{7}-\frac{7}{8}\right)=$

4 A bakery has six bags of flour. They are asked to make four loaves of bread.

The first two loaves each require two bags and a quarter of another bag.
The next loaf requires one bag and two thirds of another bag.
The final loaf requires one and a half bags of flour.

a Write a calculation, using brackets, to work out how much flour the bakery will have left after making these loaves.

b Calculate the amount of flour the bakery will have left.

Exercise 14.3

1 Calculate the following, showing your method clearly and giving your answer in its simplest form.

a $\frac{3}{8} \times 2\frac{1}{3} =$

b $5\frac{1}{6} \times 2\frac{7}{10} =$

c $3\frac{7}{8} \div \frac{7}{24} =$

2 A furlong is a measurement of distance often used in horse racing. There are approximately 200 metres in a furlong, and 8 furlongs in a mile.

Two horses, A and B, are running on a racetrack. Horse A is 1 furlong and 20 m ahead of horse B. After 4 minutes, horse B has travelled 1 and a half miles, whereas horse A has travelled 13 furlongs.

a What distance has horse A run in total? Give your answer in furlongs.

b Which horse ran the greater distance? Justify your answer.

3 Calculate $\left(2.4 \div 1\frac{3}{5}\right)^2 - \frac{3}{4} \times 1\frac{1}{5}$

15 Manipulating algebraic expressions

Exercises 15.1–15.2

1 For each of the following shapes, write an expression for the area, using brackets. Then expand the brackets and simplify your answer.

a

$t + 11$

$t + 1$

Area with brackets = ...

....

Area without brackets = ...

....

b

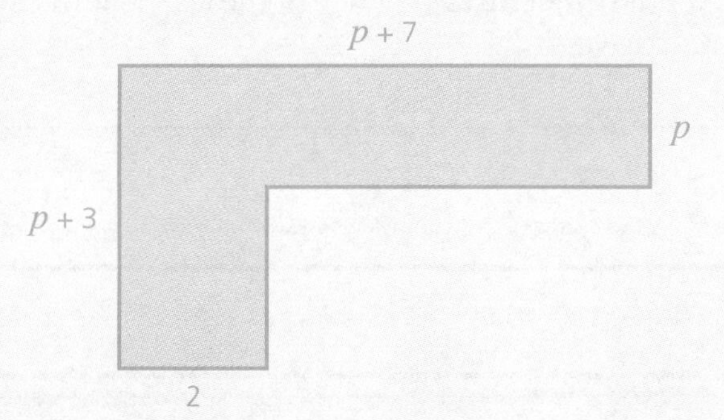

Area with brackets = ...

....

Area without brackets = ...

....

2 Expand the following. Simplify your answers fully.

a $(x+3)(x+4)$

....

....

b $(a-5)(a+4)$

c $(t-6)(t-2)$

d $(q-7)^2$

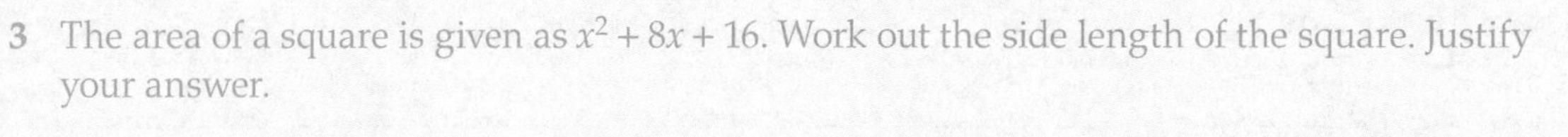

3 The area of a square is given as $x^2 + 8x + 16$. Work out the side length of the square. Justify your answer.

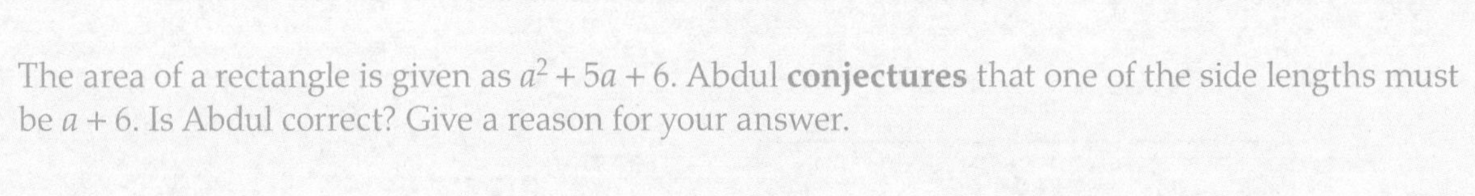

4 The area of a rectangle is given as $a^2 + 5a + 6$. Abdul **conjectures** that one of the side lengths must be $a + 6$. Is Abdul correct? Give a reason for your answer.

Exercise 15.3

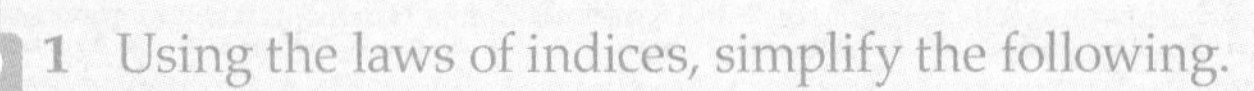

1 Using the laws of indices, simplify the following.

a $3a^6 \div a^3 =$ ……………………

b $(4c^3)^2 \div 8c^4 =$ ……………………

2 Simplify the following.

a $(w^3 \times 2w^6) \div (w \times w^2) =$ ……………………

b $(r^3)^{-4} \times r^2 \div (r^4)^{-2} =$ ……………………

3 Lucia says that $(a^3 \times 10a^2) \div (5a \times a^2)$ simplifies to $2a^4$. Lucia is incorrect.

a What mistake could Lucia have made?

b Work out the correct answer.

4 Write an expression for the area of the triangle. Give your answer in its simplest expanded form.

Exercise 15.4

1 Simplify the following expressions.

a $\frac{6x^3}{x} =$ ....

b $\frac{9y^4}{3y} =$ ....

c $\frac{15p^7q^2}{3p^2q^4} =$ ....

2 The following rectangle has an area given by the expression $20x^2 + 12x$.

The width of the rectangle is $4x$.

$4x$

The rectangle is split into four identical rectangles, as shown below.

a Find an expression for the area of one of the smaller rectangles.

b Find the length of one of the smaller rectangles.

3 Simplify the following expressions fully.

a $\frac{x^2+4x+3}{2x+6} =$

b $\frac{x^2-64}{x+8} =$

16 Combined events

Exercise 16.1

1 Amara is selecting people to be on her team. There are three girls and four boys to choose from. She picks two people at random.

a Draw a tree diagram to show all the possible outcomes.

b Calculate the probability that Amara selects two girls.

....

....

c Calculate the probability that she selects one girl and one boy, in any order.

....

....

d Is the probability of picking a boy second independent of which gender she picked first? Justify your answer.

....

....

....

2 Asif is competing in a long jump competition. He is allowed three attempts to get through to the next stage. If he jumps 6 m or further, then he gets through to the next stage and does not need to jump again. If he does not jump 6 m then he must jump again. The probability of Asif jumping 6 m or further is 0.8.

a Draw a tree diagram to represent all the possible outcomes.

b Calculate the probability that Asif gets through to the next stage on his second jump.

....

....

c Calculate the probability that Asif does not get through to the next stage.

....

....

17 Further constructions, polygons and angles

Exercise 17.1

1 Bisect the angle ABC. Leave all your construction lines.

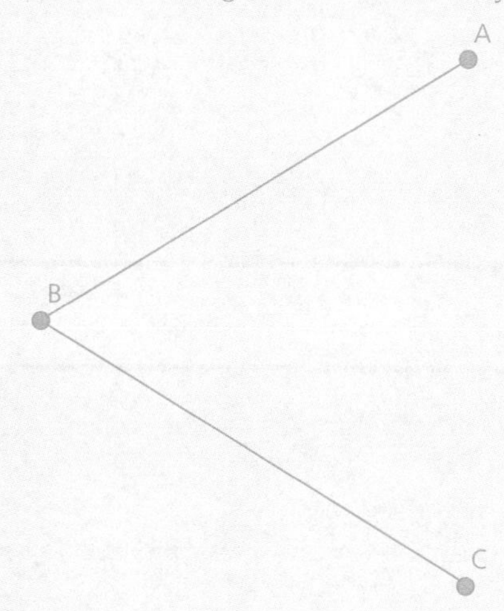

2 Construct a right-angled isosceles triangle XYZ where angle XYZ = 90°, XY = YZ and XZ = 8 cm.

3 Construct angles of the following sizes.

a 150°

b 105°

Exercise 17.2

1 The exterior angle of a regular polygon is 18°.

a Calculate the number of sides the polygon has.

....

....

b Calculate the size of each interior angle.

....

....

2 The pattern shows regular octagons tessellating with squares and a hexagon.
By highlighting its characteristics, prove that the hexagon is not regular.

....

....

....

....

....

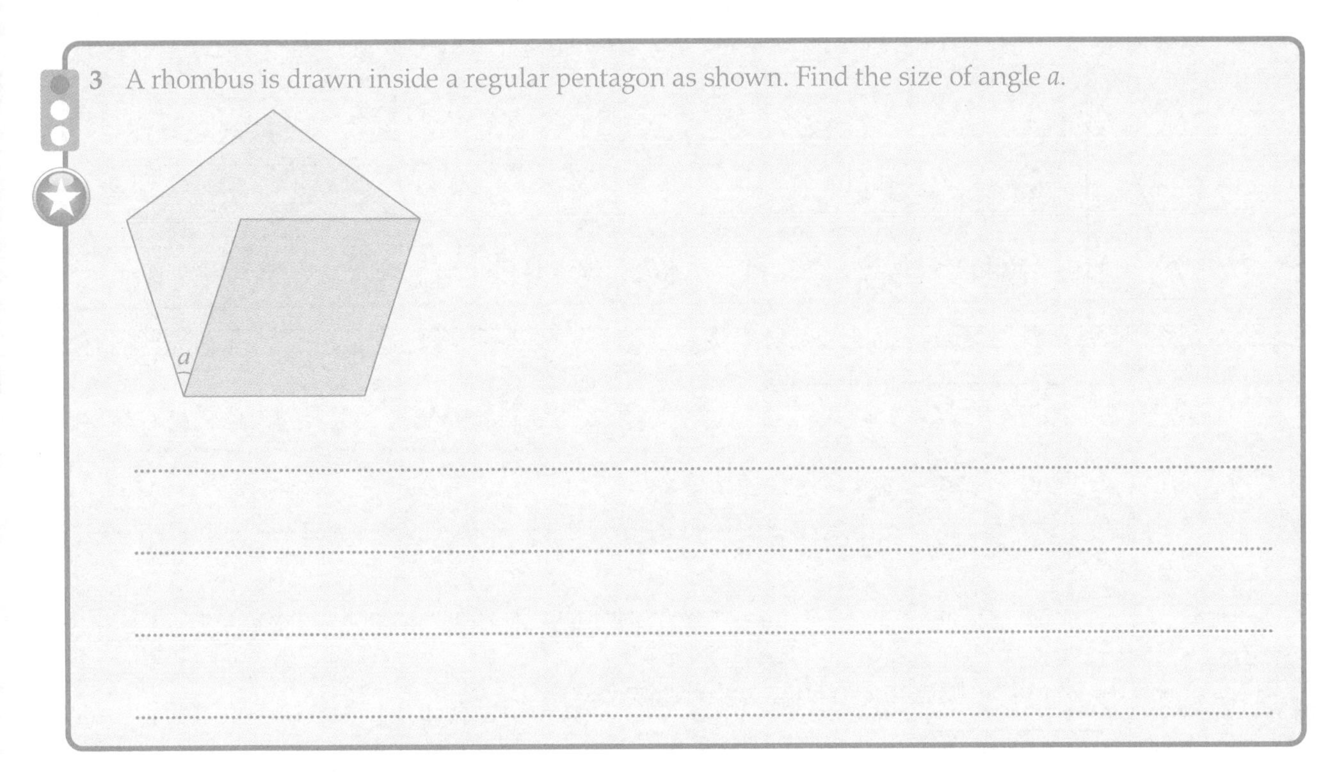

3 A rhombus is drawn inside a regular pentagon as shown. Find the size of angle a.

Exercise 17.3

1 Calculate the size of each unknown angle in these diagrams.

a

95°
a
56°

b

40°
b

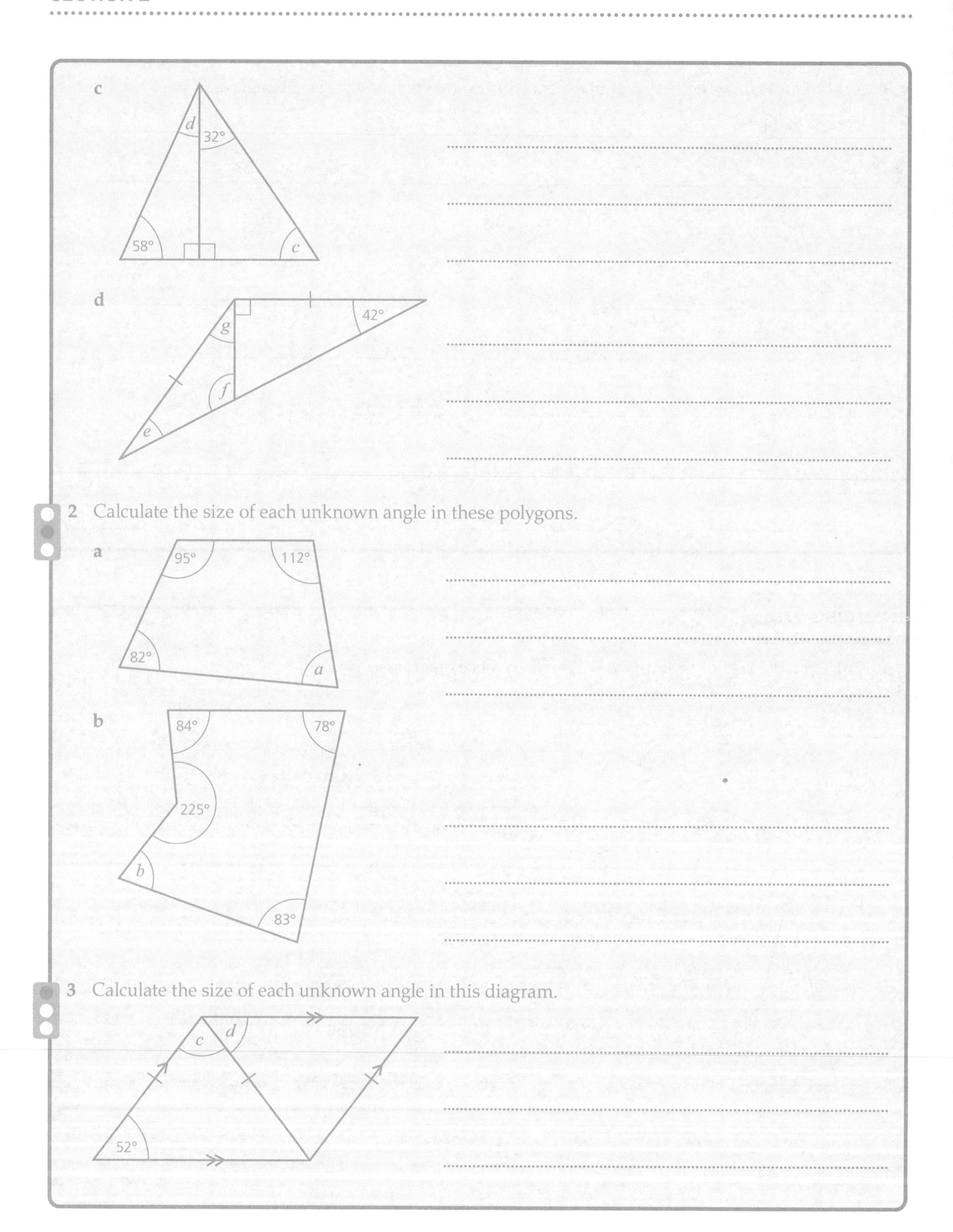

2 Calculate the size of each unknown angle in these polygons.

3 Calculate the size of each unknown angle in this diagram.

18 Further algebraic expressions and formulae

Exercise 18.1

1 A rectangle has dimensions as shown.

$x + 6$

$x - 3$

a Write an expression for

i its perimeter

....

....

ii its area.

....

....

b If $x = 8$, evaluate the expressions in part (a).

i Perimeter = ...

ii Area = ...

2 A rectangle has a side length a. The width of the rectangle is double the length, once you have added 1 to the length.

a Write an expression for the width of the rectangle.

....

b Write an expression for the perimeter of the rectangle.

....

....

c Write an expression for the area of the rectangle.

....

....

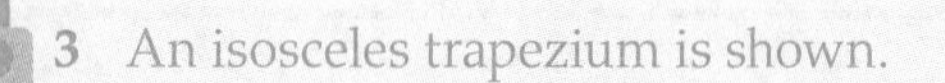

3 An isosceles trapezium is shown.

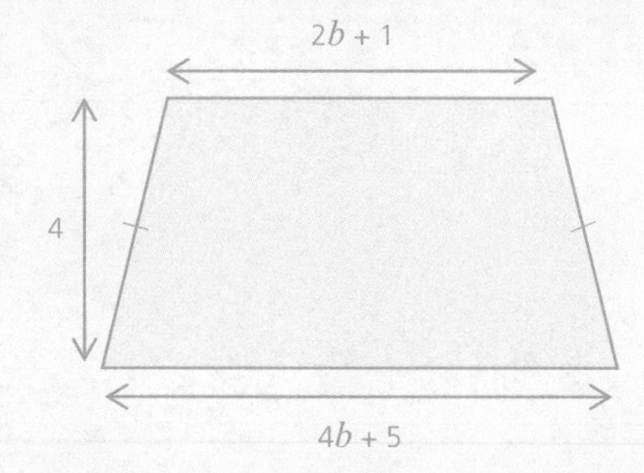

Find an expression in terms of b for the area of the trapezium.

....

....

Exercise 18.2

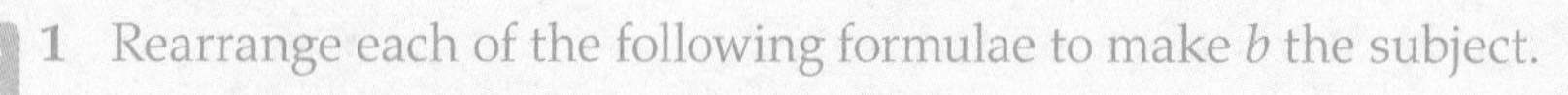

1 Rearrange each of the following formulae to make b the subject.

a $y = 2b + a$

....

....

b $p = b^2 - 11$

....

....

c $f = \frac{1}{3}\sqrt{b}$

....

....

2

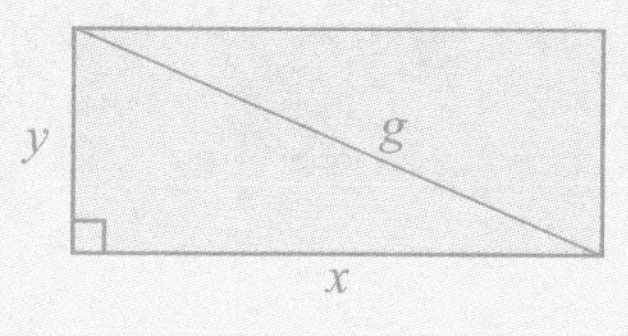

a For the rectangle shown, give a **convincing** reason why $g^2 = x^2 + y^2$.

....

....

b Rearrange the formula to make y the subject.

....

....

c Calculate the value of y if $g = 29$ and $x = 21$.

....

....

3 A shape is made from a rectangle and two quarter-circles as shown.

a Show that the area of the shape is given by the formula $A = \frac{\pi r^2}{2} + wr$.

....

....

b Rearrange the formula to make w the subject.

....

....

c Calculate the value of A if $r = 4.5$ cm and $w = 12$ cm. Give your answer correct to 2 decimal places.

....

....

....

....

19 Probability – expected and relative frequency

Exercise 19.1

1 A family of four people are using the spinner below for a game and they wonder if it is biased. They each spin the spinner several times; the results are in the table below.

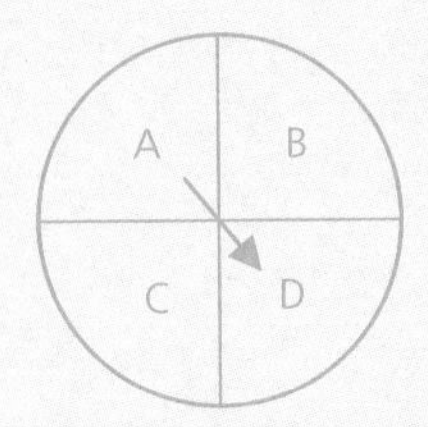

Name	Number of spins	A	B	C	D
Akanksha	40	11	12	6	11
Zola	120	28	47	17	28
Kwame	350	72	115	55	108
Safiya	200	47	73	32	48

a Whose results are the most likely to give the most reliable relative frequency? Give a reason for your answer.

....

....

b Collect all the results into a single table. Is the spinner biased or unbiased? Justify your answer.

....

....

c Calculate the relative frequency of landing on A. Give your answer correct to 2 decimal places.

....

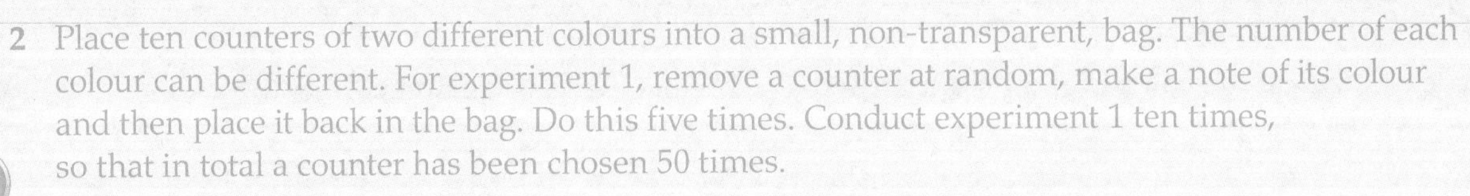

2 Place ten counters of two different colours into a small, non-transparent, bag. The number of each colour can be different. For experiment 1, remove a counter at random, make a note of its colour and then place it back in the bag. Do this five times. Conduct experiment 1 ten times, so that in total a counter has been chosen 50 times.

a Construct a table showing the relative frequency of each colour over the ten experiments.

b i On the axes below, plot a line graph for the relative frequency of each colour over the ten experiments.

Relative frequency

Number of trials

ii Describe the shape of the graphs in relation to each other.

....

....

....

....

iii Give a reason for your findings in part (ii).

....

....

....

....

20 Further algebraic equations and inequalities

Exercise 20.1

1 Solve the following equations.

a $\frac{15}{x} = 3$

....

b $\frac{64}{y} + 3 = 19$

....

c $\frac{42}{2g} - 5 = -2$

....

2 Two students are solving the following equation.

$$\frac{12x}{4x+3} = 2$$

Below are their methods.

Student 1:

1 Multiply both sides by $4x + 3$

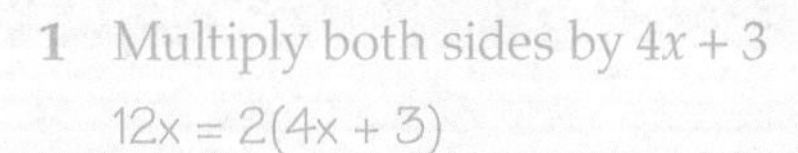

12x = 2(4x + 3)

2 Divide both sides by 2

6x = 4x + 3

3 Subtract $4x$ from both sides

2x = 3

4 Divide both sides by 2

x = 1.5

Student 2:

1 Multiply both sides by $4x + 3$

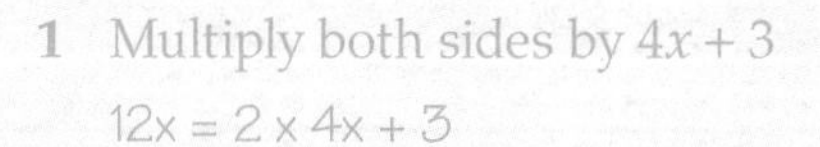

12x = 2 x 4x + 3

2 Simplify

12x = 8x + 3

3 Subtract $8x$ from both sides

4x = 3

4 Divide both sides by 4

x = 0.75

One of the students has made a mistake. Which student is it? Justify your answer.

....

....

....

....

3 Solve the following equations.

a $\frac{5x}{x+6}=3$

b $\frac{12x+4}{5x-2}=4$

c $6=\frac{2(x-5)}{3-x}$

4 A parallelogram is shown.

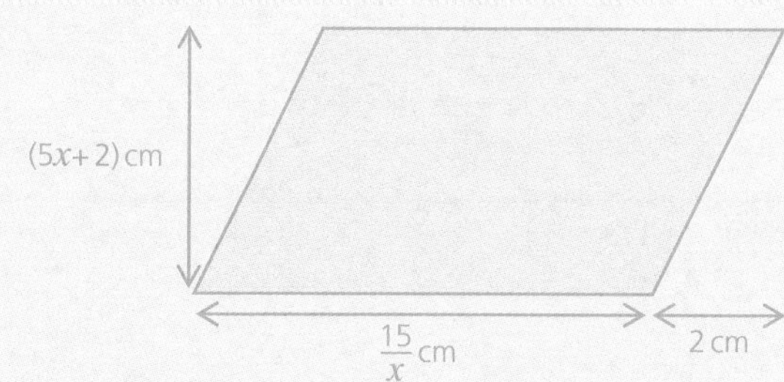

The area of the parallelogram is $125\,\text{cm}^2$.

a Calculate the height.

b Calculate the length of the sloping edge correct to 2 significant figures.

Exercise 20.2

1 Solve the following inequalities and express each answer on a number line.

a $\frac{3x}{7} \leq 6$

b $5(x - 3) < 4x + 3$

c $8 \leqslant 2x - 4 < 18$

2 Here are two function machines, A and B.

A Input → Multiply by $\frac{3}{2}$ → Add 5 → Output

Both machines have the same input.

Calculate the range of input values for which the output of A is greater than the output of B.

....

....

....

3 A gardener is designing a flower garden. He has a rectangular flower bed as shown.

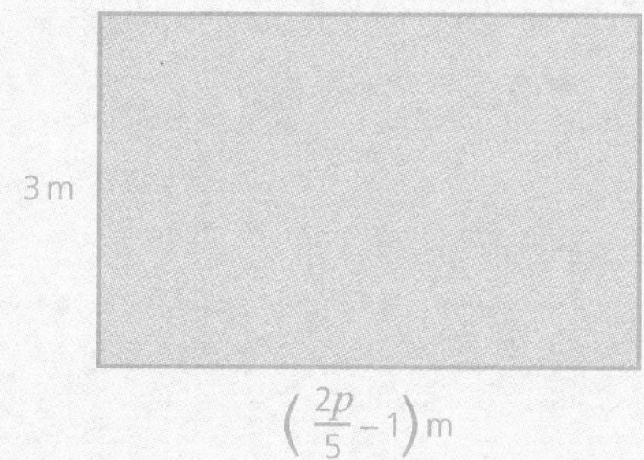

The area of the whole flower bed must be greater than $12\,m^2$ but less than or equal to $18\,m^2$.

a Find the integer values that p could be equal to.

....

....

....

b The gardener wants the perimeter of the rectangle to be greater than 15 m.

What is the smallest integer value that p could be?

....

....

....

SECTION 3

21 Linear and quadratic sequences

Exercise 21.1

1 In each of the following sequences, let the first term (t_1) be 4. Generate the first five terms using the term-to-term rule in each case.

a Double the previous term and subtract 1 ………… , ………… , ………… , ………… , …………

b Square the previous term and subtract 10 ………… , ………… , ………… , ………… , …………

c Divide the previous term by 2 and add 4 ………… , ………… , ………… , ………… , …………

2 In each of the following linear sequences, calculate the missing terms.

a ………… , 4, ………… , 12, …………

b −1, ………… , ………… , 8, …………

c ………… , 6, ………… , −4, …………

3 Below is a sequence of five numbers. The first term is given as a.

a, ………… , ………… , ………… , …………

The term-to-term rule is double the previous term and add 5.

a Calculate an expression for the third term.

……………………………………………………………………………………………………

……………………………………………………………………………………………………

b The sum of the first five terms is 223. Calculate the value of a.

……………………………………………………………………………………………………

……………………………………………………………………………………………………

……………………………………………………………………………………………………

c Calculate the value of the fifth term.

……………………………………………………………………………………………………

……………………………………………………………………………………………………

Exercise 21.2

1 For each of the following sequences, the terms and their position within the sequence are given.

a

Position	1	2	3	4	5	n
Term	4	6	8	10	12	

i Calculate the term-to-term rule.

....

ii Find the nth term rule.

....

....

iii Calculate the 50th term.

....

....

b

Position	1	2	3	4	5	n
Term	10	11	12	13	14	

i Calculate the term-to-term rule.

....

ii Find the nth term rule.

....

....

iii Calculate the 20th term.

....

....

c

Position	1	2	3	4	5	n
Term	8	5	2	–1	–4	

i Calculate the term-to-term rule.

....

ii Find the *n*th term rule.

....

....

iii Calculate the 100th term.

....

....

Exercise 21.3

1 Calculate the next two terms of each sequence. Justify your answers.

a 1, 4, 9, 16, 25, ... , ...

....

....

b 5, 7, 10, 14, 19, ... , ...

....

....

c 8, 7, 5, 2, −2, ... , ...

....

....

2 Calculate the first five terms of the sequence given by the following *n*th term rule.

$t_n = n^2 - 3$

....

....

....

.........................., ..., ..., ..., ...

3 The first five terms in the sequence of square numbers are 1, 4, 9, 16, 25 and the rule for the nth term is $t_n = n^2$.

Compare the following sequence to the square number sequence above.

−4, −1, 4, 11, 20

a Deduce the nth term rule.

....

....

b Calculate the 10th term.

....

....

4 A sequence of shapes is constructed using small cubes.

Position 1 2 3

a Find the **general** rule for the nth term of the number of cubes in each shape of the sequence.

....

....

b How many cubes will be needed to construct the 10th shape in the sequence?

....

....

c Can shape 25 be constructed using 620 small cubes? Justify your answer.

....

....

....

....

22 Compound percentages

Exercise 22.1

1 a Which of the following multipliers is equivalent to a 6% increase? Justify your answer.

$\times 1.6$ $\times 1.06$ $\times 6$ $\times 106$

....

....

b Which of the following multipliers is equivalent to a 23% decrease? Justify your answer.

$\times 1.23$ $\times 1.77$ $\times 77$ $\times 0.77$

....

....

2 Owing to conservation efforts over the past 10 years, the number of tigers in the wild is increasing by 3% per year. In 2019 there were recorded to be 3900 tigers in the wild.
If this rate continues, in what year will there be over 5000 tigers in the wild?

....

....

....

....

3 Abdul invests \$6000 in a bank with a compound interest rate of 4% per annum. At the end of each year he has to pay 20% tax on the interest added during that year. After 5 years, Abdul wants to buy a car valued at \$7000.
Will Abdul have enough money to buy the car? If so, how much will he have left over? If not, how much more money would he need to save?

....

....

....

....

Scale and area factors of enlargement

Exercise 23.1

1 The table below shows some scale factors of enlargement and their corresponding area factors of enlargement. Complete the table with the missing factors of enlargement.

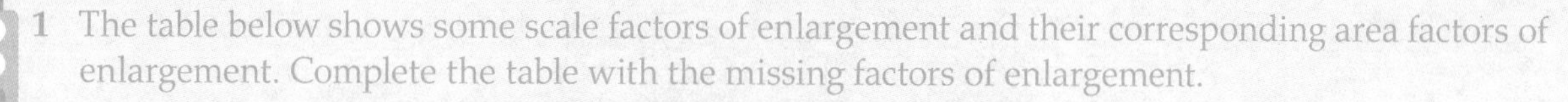

Scale factor of enlargement	Area factor of enlargement
3	
10	
	256
	49
x	

2 Two rectangles are shown below. One is an enlargement of the other by a scale factor of 3.

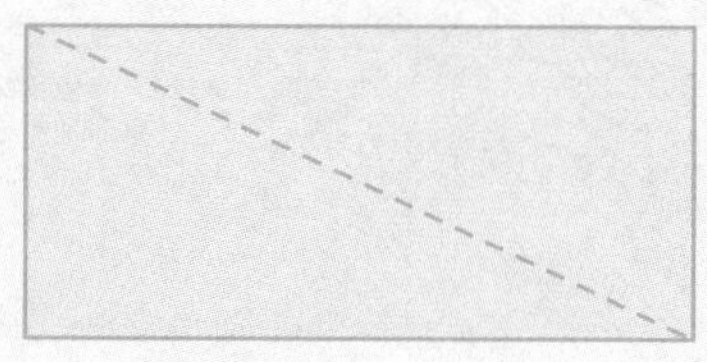

a Calculate the area of the enlarged rectangle.

....

....

b Calculate the length of the diagonal of the enlarged rectangle.

....

....

3 Two trapezia, A and B, are shown below. B is an enlargement of A. The area of trapezium B is $256\,cm^2$.

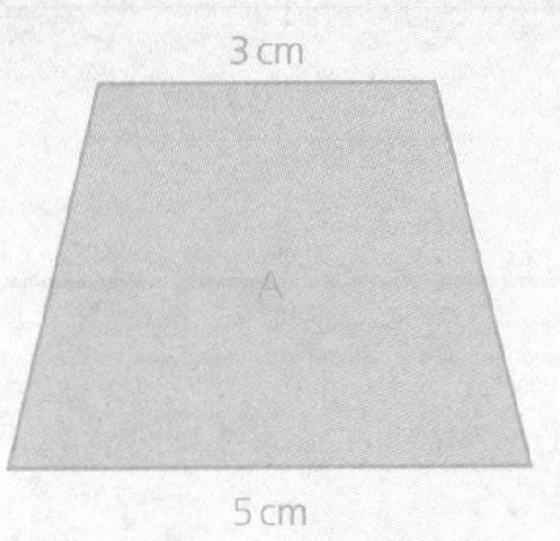

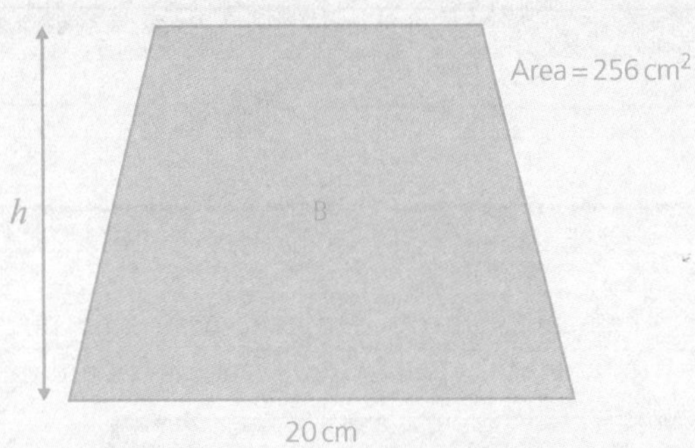

a Calculate the height of trapezium B, labelled h.

....

....

....

b Calculate the area of trapezium A.

....

....

c Work out the ratio of area of A : area of B. Give your answer in its simplest form.

....

....

24 Functions and their representation

Exercise 24.1

1 Complete the following table for the function $y = 2x^2 - 5$. If there are two possible values, include both.

Input	Output
–4	
0	
$\frac{1}{2}$	
	13

2 Complete the mapping diagram for the graph below.

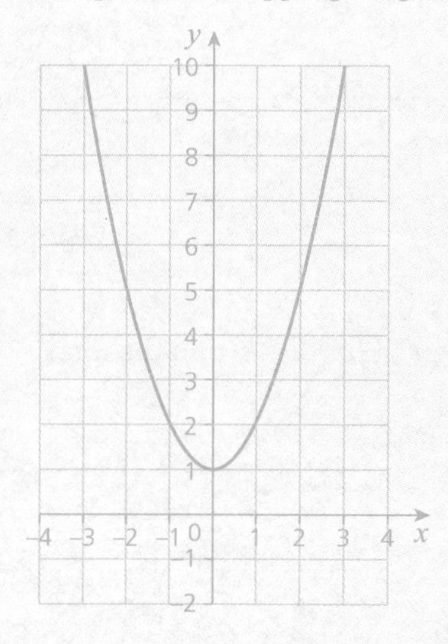

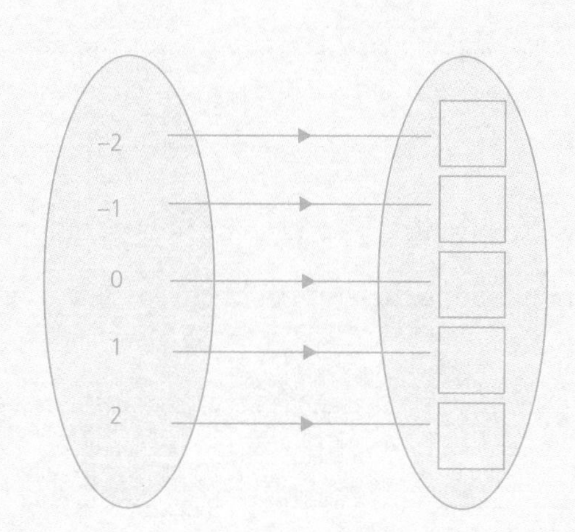

3 An equation is given as $y = (4x)^2$.

A student states that the function machine for this equation is

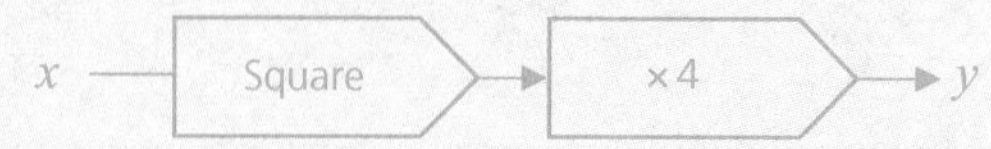

a Explain why this function machine is incorrect for $y = (4x)^2$.

....

....

b What is the equation for the function machine above?

....

....

c A student uses the function machine and says that if the output is positive the input must also be positive. Comment on this statement.

....

....

....

Exercise 24.2

1 Three of the four equations below represent the same straight line. By considering their special properties, circle the odd one out and give a reasoned answer for your choice.

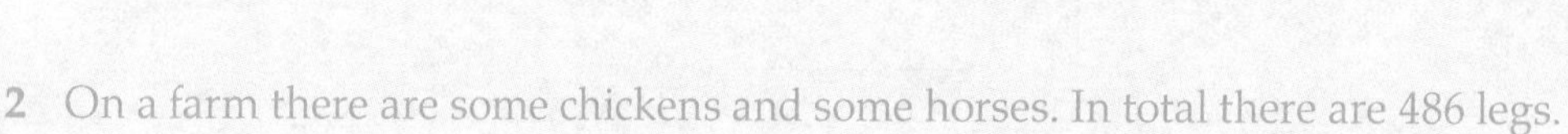

$3x = y + 5$ $\quad$ $y = 3x + 5$ $\quad$ $y - 5 = 3x$ $\quad$ $y - 3x - 5 = 0$

....

....

....

2 On a farm there are some chickens and some horses. In total there are 486 legs.

a Write an equation linking the number of chickens (c), the number of horses (h) and the total number of legs.

....

....

....

b If the number of chickens is greater than 40, work out a possible combination of chickens and horses on the farm.

....

....

....

c Explain why there cannot be 150 chickens.

3 A football manager is buying his team a new football kit. A football shirt costs \$32 and a pair of shorts costs \$25. The manager spends a total of \$1012.

a Write an equation linking the total number of shirts (f) the manager bought, the total number of pairs of shorts (s) the manager bought and the total amount he spent.

b If the manager bought 16 shirts, how many pairs of shorts did he buy?

c Explain why the manager must have bought 16 shirts.

25 Coordinates and straight-line segments

Exercise 25.1

1 The straight-line segment AB has end coordinates A(2, 2) and B(6, 22). Calculate the coordinates of the following points.

a M, the midpoint of AB

....

....

b X, if X divides AB in the ratio 1 : 3

....

....

c C, if B is the midpoint of AC

....

....

2 A straight-line segment is drawn, where O is at the origin and points A, B and C are equally spaced along its length as shown.

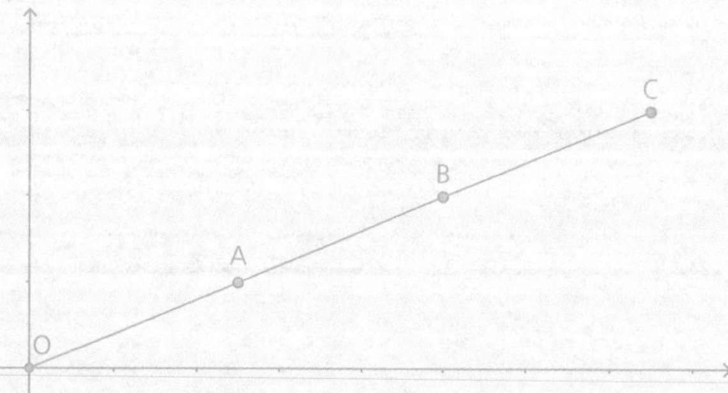

The coordinates of C are (15, 6). Find the coordinates of A and B.

....

....

....

A = ... B = ...

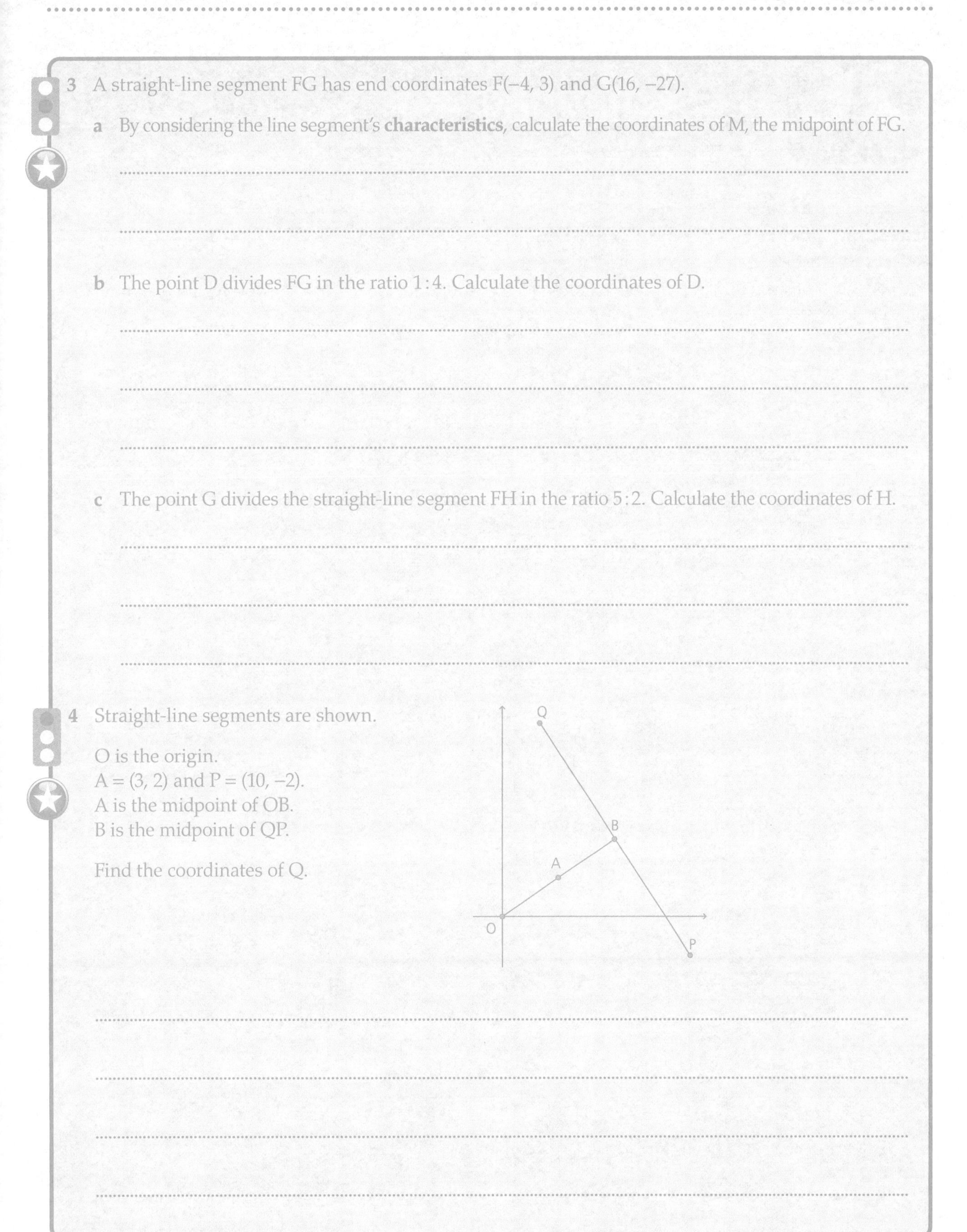

3 A straight-line segment FG has end coordinates F(–4, 3) and G(16, –27).

a By considering the line segment's **characteristics**, calculate the coordinates of M, the midpoint of FG.

....

....

b The point D divides FG in the ratio 1 : 4. Calculate the coordinates of D.

....

....

....

c The point G divides the straight-line segment FH in the ratio 5 : 2. Calculate the coordinates of H.

....

....

....

4 Straight-line segments are shown.

O is the origin.
A = (3, 2) and P = (10, –2).
A is the midpoint of OB.
B is the midpoint of QP.

Find the coordinates of Q.

....

....

....

....

26 Estimating surds

Exercise 26.1

1 For each of the following, circle which of the given answers are definitely incorrect. Justify your choices.

a $\sqrt{86} =$ 8.7 10.3 9.3 7.9

....

....

b $\sqrt{150} =$ 11.7 12.2 10.9 13.1

....

....

c $\sqrt[3]{129} =$ 4.3 4.8 6.4 5.1

....

....

2 A rectangle has dimensions as shown.

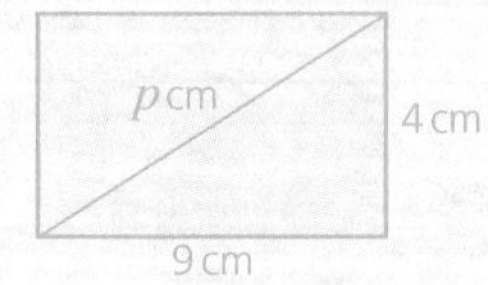

a Prove that $p = \sqrt{97}$

....

....

b Write the value of p in the form $a < p < b$, where a and b are integers.

....

....

3 A rectangle has dimensions as shown.

Area = $100\ cm^2$

$(x - 2)$ cm

$(x + 2)$ cm

a Write the value of x in surd form.

....

b Write the value of x in the form $a < x < b$, where a and b are the closest integer values to x.

....

....

Linear functions and solving simultaneous linear equations

Exercise 27.1

1 Three functions are given below. Circle the quadratic function(s).

$y = x^2 - 3$ $\quad$ $y - x^2 = 5$ $\quad$ $y = \frac{1}{2}x + 10$

2 **a** Complete the table of values for the function $y = \frac{1}{2}x + 5$.

x	−4	−2	0	2	4
y					

b Plot the graph of the function $y = \frac{1}{2}x + 5$ on the axes below.

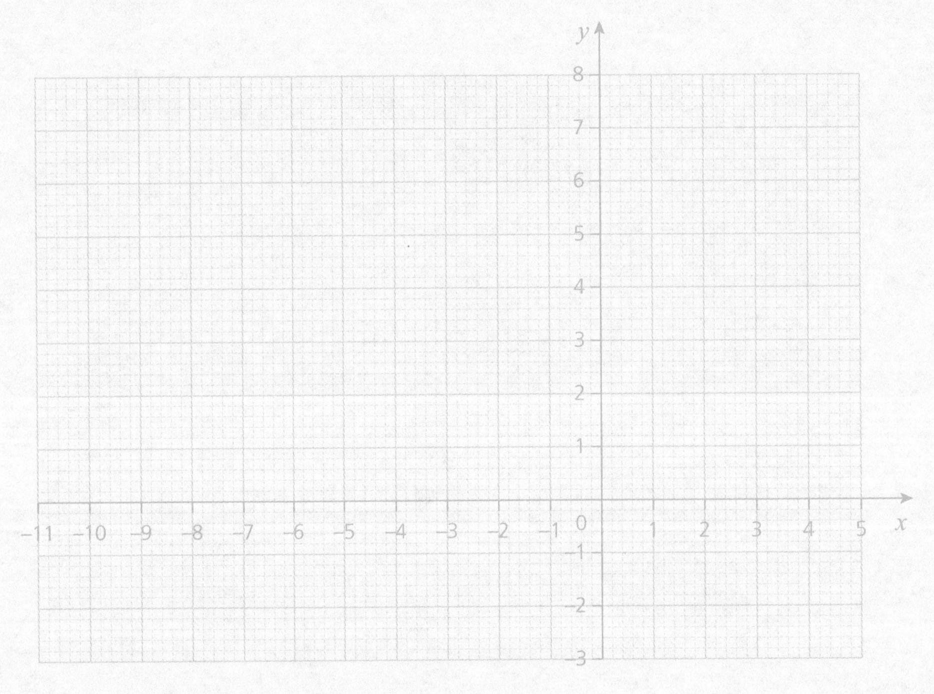

3 a Complete the table of values for the function $y = x^2 - 2$.

x	−3	−2	−1	0	1	2	3
y		2					7

b Plot the graph of the function $y = x^2 - 2$ on the axes below.

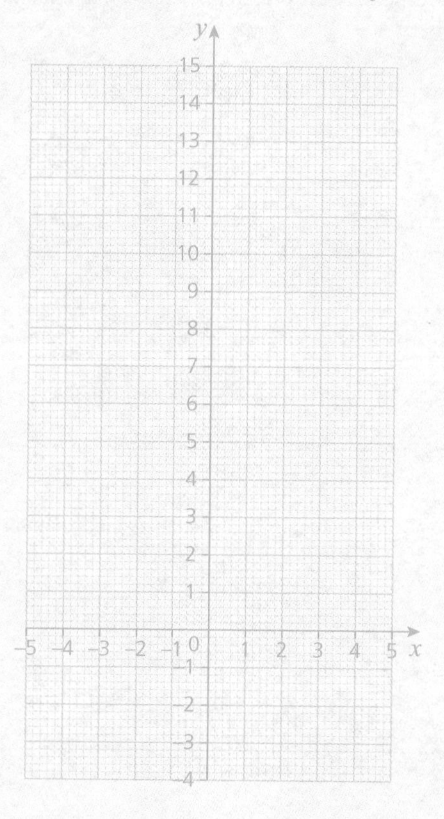

4 a Deduce the coordinates where the line $2x + 3y = -6$ crosses the y-axis.

....

....

b Deduce the coordinates where the line $2x + 3y = -6$ crosses the x-axis.

....

....

c Prove that the point (4.5, −5) lies on the line $2x + 3y = -6$.

....

....

Exercise 27.2

1 Calculate the gradient of the straight line passing through each of these pairs of points.

a (2, 4) and (4, 8)

....

....

b (2, 8) and (5, 9)

....

....

c (5, 4) and (4, 7)

....

....

d (–2, 13) and (2, 12)

....

....

2 Calculate the gradient and the y-intercept of the line below.

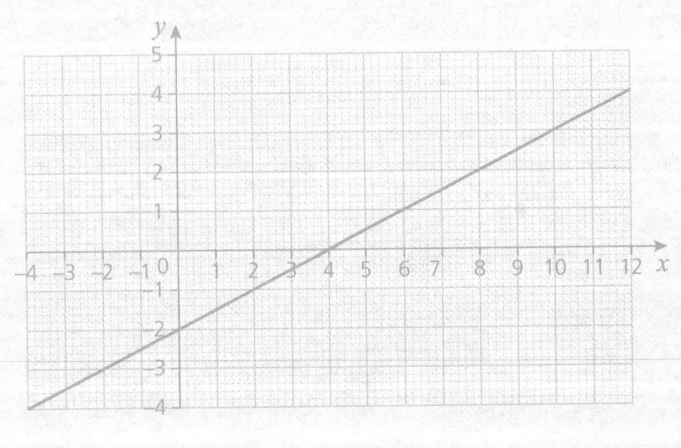

....

....

Gradient = ... y-intercept = ...

3 For the straight lines represented by each of the equations below, find the value of the gradient and the y-intercept.

a $y = 3x - 6$

Gradient = ... y-intercept = ...

b $y - x = -8$

Gradient = ... y-intercept = ...

c $y + \frac{1}{2}x = -4$

Gradient = ... y-intercept = ...

4 A straight line is shown.

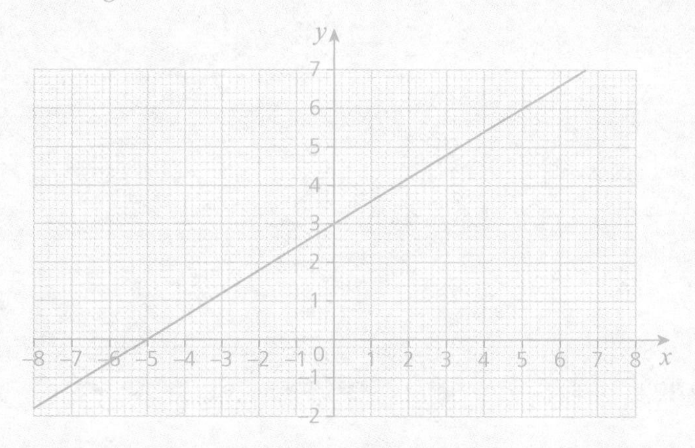

a Calculate the equation of the straight line, giving your answer in the general form $y = mx + c$.

....

....

b The line is reflected in the y-axis. Calculate the equation of the reflected line, giving your answer in the form $y = mx + c$.

....

....

c Compare the equations of the two lines from parts (a) and (b). Comment on what you notice.

....

....

....

Exercises 27.3–27.4

1 Here is the graph of $y = 6 - 2x$.

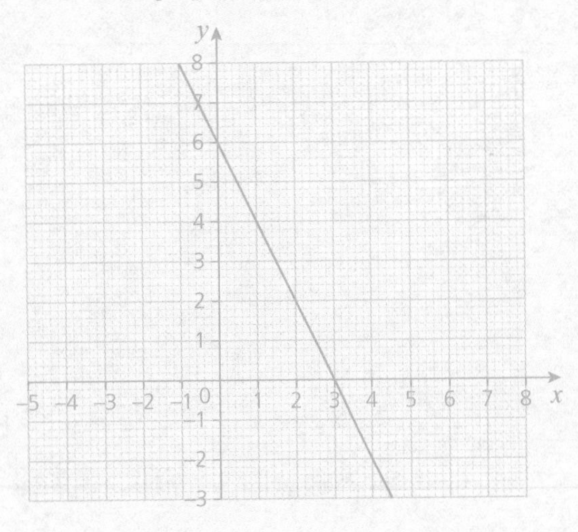

a Plot the graph of $y = \frac{1}{2}x + 1$ on the same axes.

b Using the graphs, find the coordinates of their point of intersection.

(............................ , ...)

2 For the pair of equations below, draw the two straight lines on the pair of axes provided. Then use the coordinates of the point of intersection to find an approximate solution to the simultaneous equations.

$y = \frac{1}{2}x - 3$ and $y = -3x + 2$

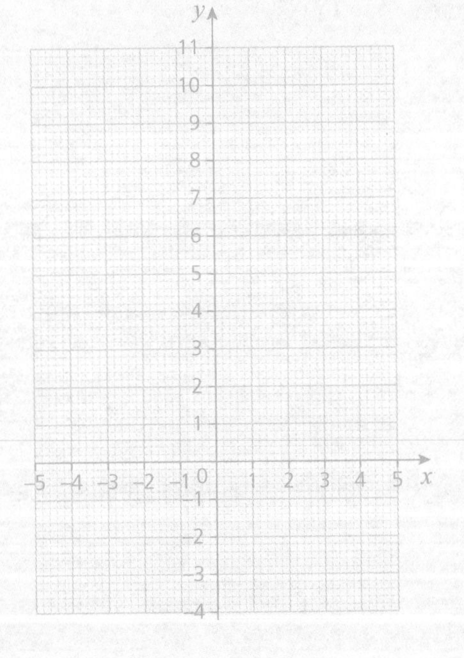

$x =$ .... $y =$ ....

3 Using the pair of axes below, or otherwise, solve the simultaneous equations.

$4x + y = 13$

$2x + y = 7$

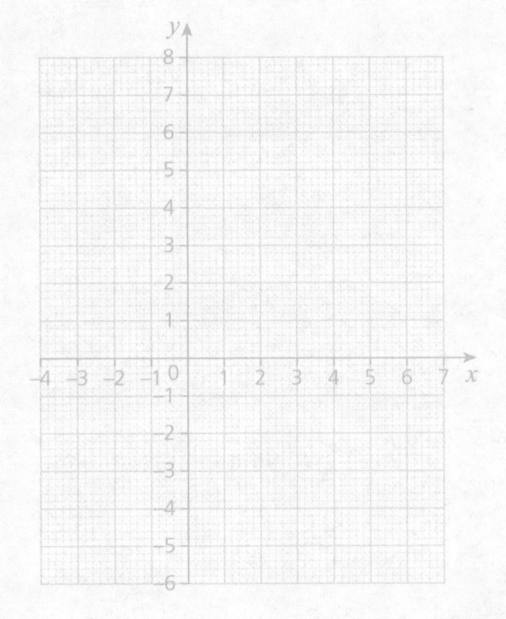

....

....

....

Exercises 27.5–27.6

1 Solve the following pairs of simultaneous equations by elimination.

a $3x + 2y = 24$

$x + 2y = 12$

....

....

....

....

b $4x - 3y = -3$

$4x + 5y = -27$

....

....

....

....

2 Solve the following pair of simultaneous equations by elimination.

$3w + 2v = 10$

$7w - v = 29$

....

....

....

....

3 Two groups purchase some tickets for a show at the theatre.
The first group buys four adult tickets and two child tickets for $29.
The second group buys three adult tickets and four child tickets for $28.

a Calculate the price of one child ticket.

....

....

....

....

b Calculate the total cost of two adult tickets and three child tickets.

....

....

....

28 Bearings and scale drawings

Exercise 28.1

1 A car starts at point A. It travels a distance of 6 km on a bearing of 120° to point B. From B it travels 4 km on a bearing of 330° to point C.

a Draw a diagram to show these bearings and journeys.
Use the scale = 1 cm : 1 km

b What are the distance and bearing of A from C?

....

....

2 The nearest petrol station is 3 km away from the house, labelled H, on a bearing of 072°. Use the scale 1 cm = 1 km. Mark on the map the location of the petrol station and label this point P.

H•

3 The map extract shows a part of Malaysia and Singapore.
The scale of the map is 1 : 4 000 000.

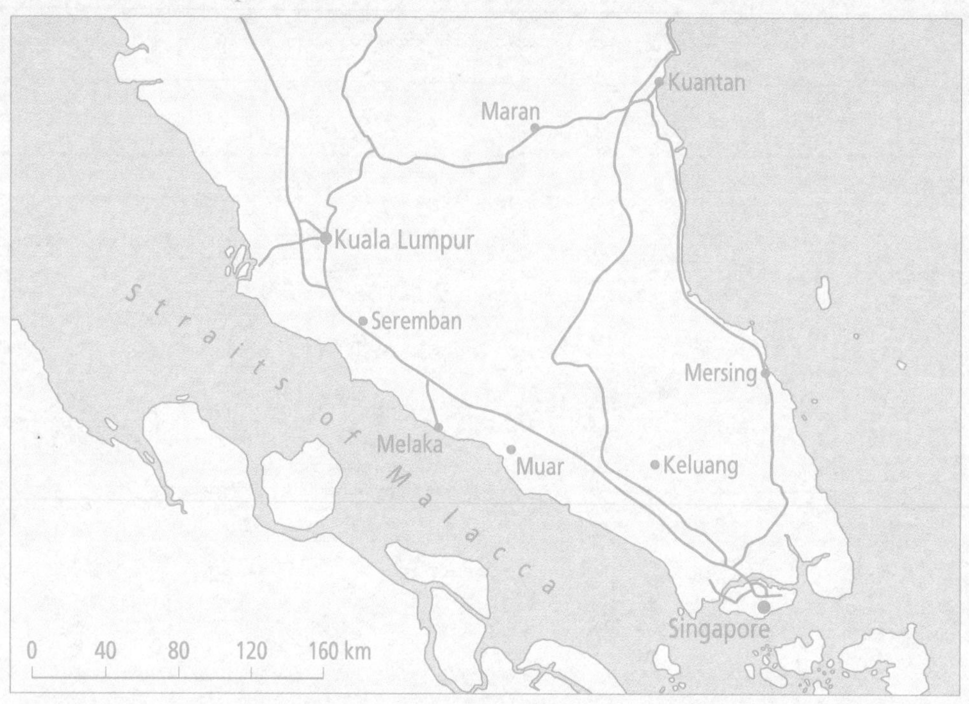

a A tourist travels from Singapore to Kuala Lumpur and then on to Kuantan. By measuring the map with a ruler, calculate the distance in km that the tourist travels in total. Assume that they travel in a straight line each time.

....

....

b What is the bearing from:

i Singapore to Kuala Lumpur?

....

ii Kuala Lumpur to Kuantan?

....

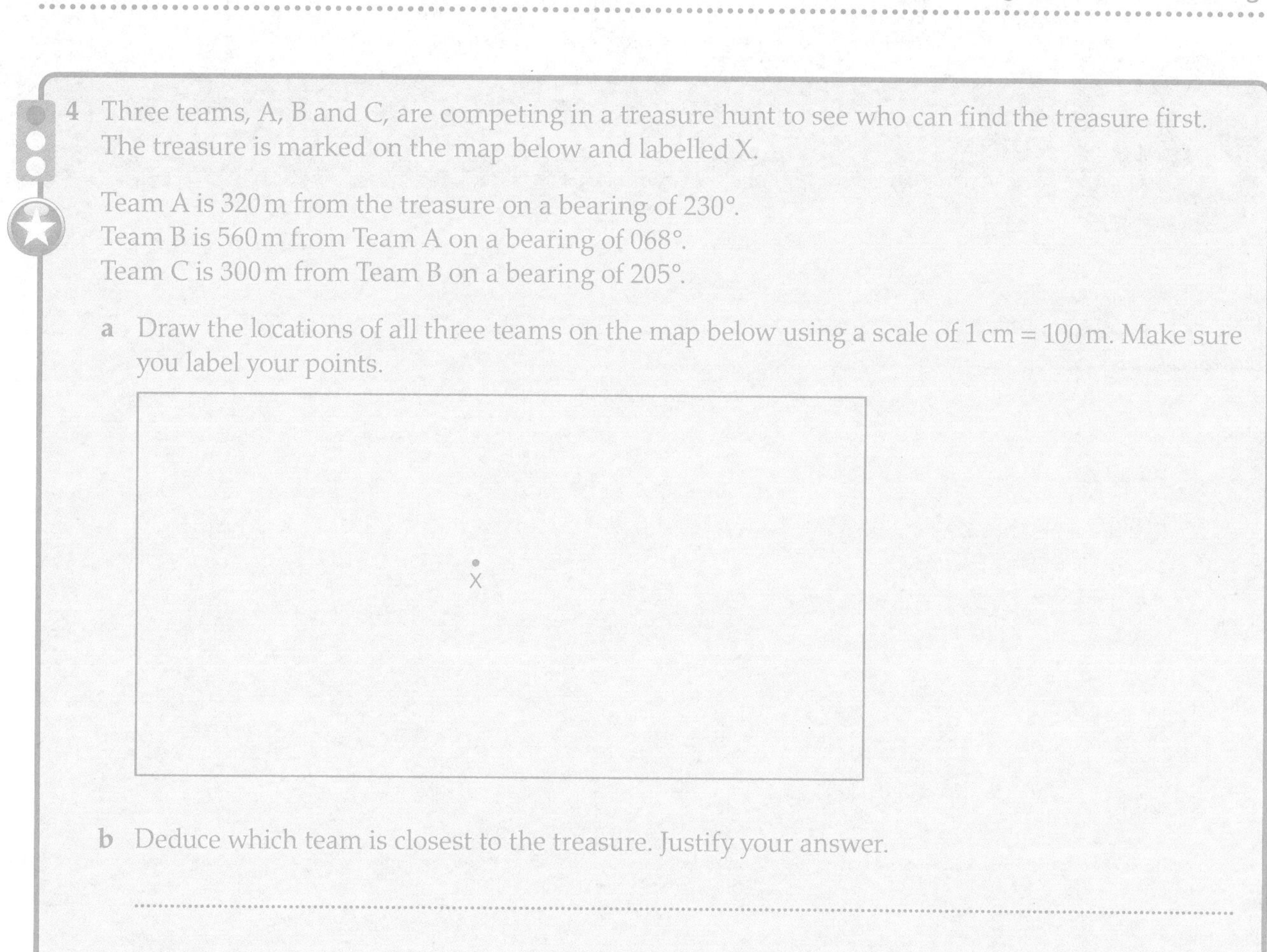

4 Three teams, A, B and C, are competing in a treasure hunt to see who can find the treasure first. The treasure is marked on the map below and labelled X.

Team A is 320 m from the treasure on a bearing of 230°.
Team B is 560 m from Team A on a bearing of 068°.
Team C is 300 m from Team B on a bearing of 205°.

a Draw the locations of all three teams on the map below using a scale of 1 cm = 100 m. Make sure you label your points.

b Deduce which team is closest to the treasure. Justify your answer.

....

....

29 Direct and inverse proportion

Exercise 29.1

1 Three footballers are practising taking a penalty. Player A takes a total of 60 penalties, and scores 42 of them. Player B takes a total of 150 penalties and scores 80 of them. Player C takes a total of 76 penalties and scores 54 of them.

a What is each player's success rate as a ratio? Give your answers in their simplest form.

....

....

Player A = ... Player B = ... Player C = ...

b What is each player's success rate as a percentage?

....

....

Player A = ... Player B = ... Player C = ...

c Which player has the best success rate? ....

2 \$6000 is divided between three people in the ratio 2 : 3 : 5. How much does each person receive?

....

....

3 Louis and Matteo are practising their dart throwing. Louis' ratio of hitting to missing is 7 : 1. Matteo throws the dart 90 times and misses 10 times.

a Matteo states that he has better accuracy than Louis. Show why Matteo is correct.

....

....

b Matteo also states that because his accuracy is better, he hit the dartboard more times than Louis did. Give a reasoned explanation showing why this might not be true.

....

....

4 The ratio of $a:b = 5:2$.
If $a + b = 56$, calculate the value of $a - b$.

Exercise 29.2

1 The ratio of boys to girls at an extra-curricular club is 3:5. At 4 p.m., four girls leave the club to go home. The ratio of the number of boys to girls at the club is now 5:7. Work out the number of boys at the extra-curricular club.

2 In a choir, the ratio of boys to girls is 2:5. There are 36 more girls than boys. Work out the total number of children in the choir.

3 The ratio of the angles in a triangle is 1:2:3. By considering its characteristics, identify what type of triangle it is. Justify your answer.

4 In a cake sale
- chocolate cakes : vanilla cakes = 3 : 1
- vanilla cupcakes : large vanilla cakes = 5 : 1.

There are three large vanilla cakes.

How many cakes are there in total?

5 The ratio of cats : dogs = 2 : 3.
The ratio of dogs : fish = 9 : 5.

What is the ratio of cats : fish? Give your answer in its simplest form.

Exercises 29.3–29.4

1 Zola walks 10 miles in four hours. How long would it take her to walk 15 miles?

2 I buy five chocolate bars for $3.20. How much would it cost to buy 12 chocolate bars?

3 It takes three builders five days to build a wall. How long would it have taken two builders?

4 In July 2020, $1 was worth €0.85. At the same time €1 was worth £0.91.

a How many euros (€) was $2100 worth?

b How many dollars ($) was €2100 worth?

c How many pounds (£) was €320 worth?

d Using the exchange rates above, how much was $1500 worth in pounds?

5 A takeaway restaurant offers delivery.

TAKEAWAY DELIVERY SERVICE

Cost of delivery is $2.50 plus $0.40 for every kilometre.

a A customer lives 4.5 km away from the restaurant. How much will delivery cost her?

b The restaurant charges a different customer $3.78 for delivery. How far away from the restaurant does this customer live?

Compound measures and graphs

Exercise 30.1

1 This graph shows the speed of a car during a journey.

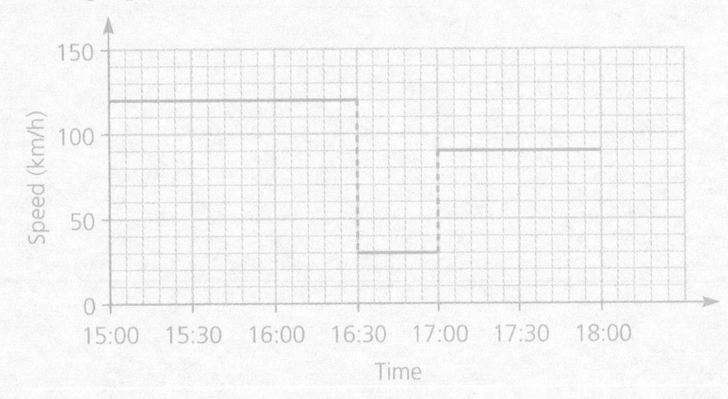

a What time did the car set off on its journey? ....

b How fast was the car travelling at 15:30? ....

c For how long did the car travel at 90km/h? ....

d The car was travelling behind a slow lorry for part of the journey. Between what times was this?

....

e How far did the car travel between 15:00 and 16:30? ....

f How far did the car travel on the whole journey? ....

2 Here is a description of a woman's run:
- she sets off at 09:00
- she runs at a constant speed of 8 km/h for the first hour
- then she stops and rests for 15 minutes
- after her rest she runs at a constant speed of 11 km/h for 20 minutes.

Plot a speed–time graph to show the woman's run.

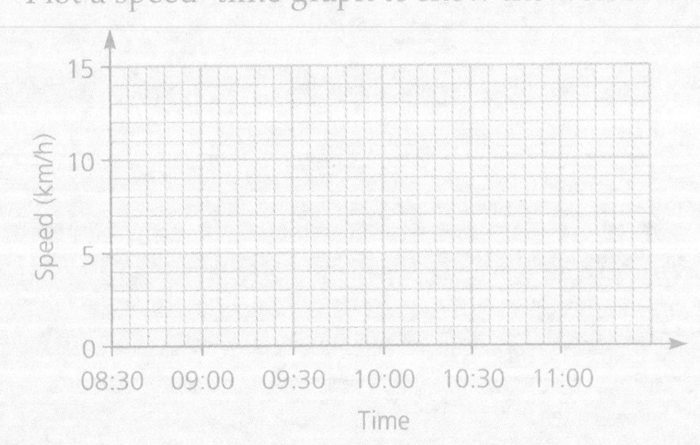

3 The distance–time graph shows Ekon's walk to meet a friend.

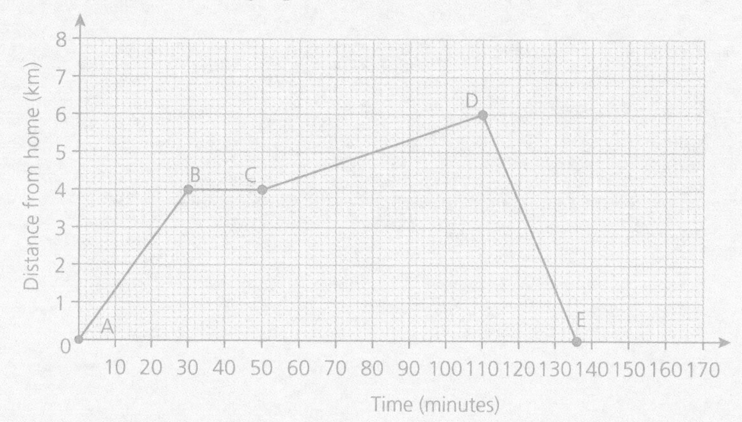

a What speed is Ekon walking at from A to B? Give your answer in km/h.

....

....

b When Ekon meets his friend, they stop and talk. Between which points on the graph is this? Justify your answer.

....

....

c Ekon borrows a friend's bike to travel home. What speed is he going when he is travelling back home? Give your answer in km/h.

....

....

Exercise 30.2

1 Two car rental companies are advertising their prices for renting a car.

Company A: $80 for the first day, $25 per extra day.
Company B: $60 for the first day, $30 per extra day.

This graph shows the costs of renting a car with both companies.

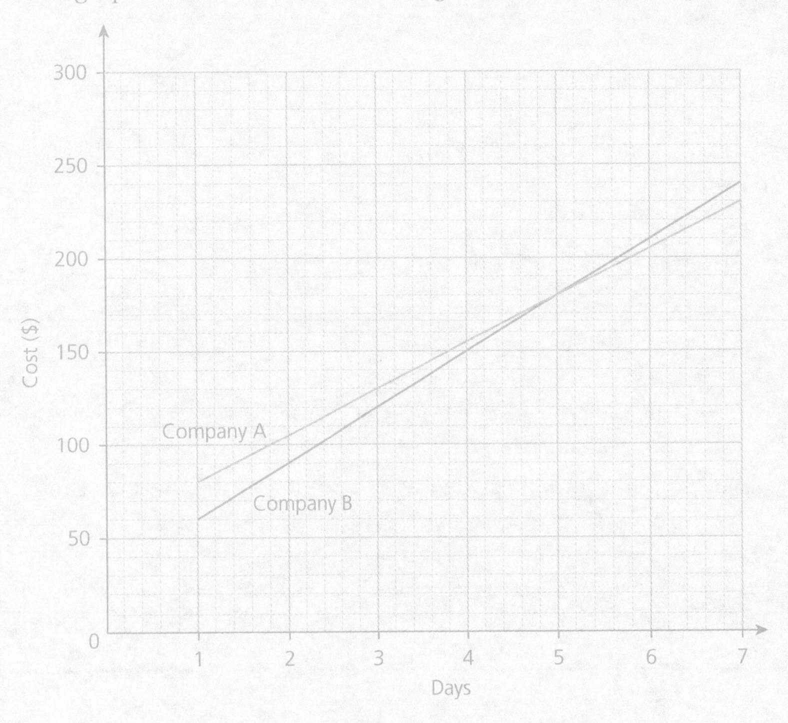

a Michelle wants to rent a car for three days. Which company would be cheaper? Justify your answer.

....

....

b After how many days does it become cheaper to rent with Company A? Justify your answer.

....

....

2 Twin babies are washed in a small bath. This graph shows the water level (in centimetres) in the bath over time.

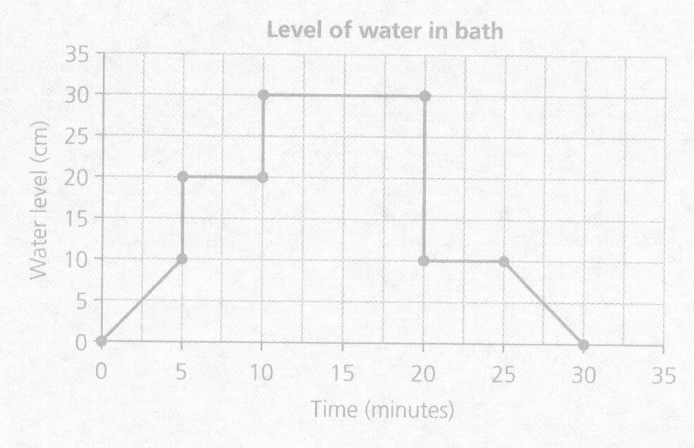

a Give a possible explanation for the shape of the graph during the first 5 minutes.

....

....

b Give an explanation for the shape of the graph at 5 minutes.

....

....

c Give an explanation for the shape of the graph at 10 minutes.

....

....

d For how long were both babies in the bath together?

....

....

e Give an explanation for the shape of the graph at 20 minutes.

....

....

The Publishers would like to thank the following for permission to reproduce copyright material.

Acknowledgements

Cambridge International copyright material in this publication is reproduced under licence and remains the intellectual property of Cambridge Assessment International Education.

Extra information on using the workbooks is accessible online via:
www.hoddereducation.com/workbook-info

Hachette UK's policy is to use papers that are natural, renewable and recyclable products and made from wood grown in well-managed forests and other controlled sources. The logging and manufacturing processes are expected to conform to the environmental regulations of the country of origin.

Orders: please contact Hachette UK Distribution, Hely Hutchinson Centre, Milton Road, Didcot, Oxfordshire, OX11 7HH. Telephone: +44 (0)1235 827827. Email education@hachette.co.uk
Lines are open from 9 a.m. to 5 p.m., Monday to Friday. You can also order through our website: www.hoddereducation.com

ISBN: 978 1 398 30130 6

First published in 2012
This edition published in 2021 by
Hodder Education,
An Hachette UK Company
Carmelite House
50 Victoria Embankment
London EC4Y 0DZ

www.hoddereducation.co.uk

Impression number 10 9 8 7 6 5 4 3 2 1

Year 2025 2024 2023 2022 2021

Cover photo © tiverylucky - stock.adobe.com

Illustrations by Integra Software Services Pvt. Ltd., Pondicherry, India

Typeset in 12/14pt Palatino LT Std by Integra Software Services Pvt. Ltd., Pondicherry, India

Printed in the UK

A catalogue record for this title is available from the British Library.

Cambridge **checkpoint**

Lower Secondary
Mathematics
WORKBOOK

9

Practise and consolidate knowledge gained from the Student's book with this write-in workbook full of corresponding learning activities.

- Save time when planning with ready-made homework or extension exercises.
- Reinforce students' understanding of key mathematical concepts with varied question types, knowledge tests and the use of ICT.
- Challenge students with extra practice activities to encourage regular self-assessment.

For over 25 years we have been trusted by Cambridge schools around the world to provide quality support for teaching and learning. For this reason we have been selected by Cambridge Assessment International Education as an official publisher of endorsed material for their syllabuses.

Working for over 25 YEARS WITH Cambridge Assessment International Education

- ✓ Provides learner support as part of a set of resources for the Cambridge Lower Secondary Mathematics curriculum framework (0862) from 2020
- ✓ Has passed Cambridge International's rigorous quality-assurance process
- ✓ Developed by subject experts
- ✓ For Cambridge schools worldwide

This series includes eBooks and teacher support.

Visit www.hoddereducation.com/boost for more information.

Registered Cambridge International Schools benefit from high-quality programmes, assessments and a wide range of support so that teachers can effectively deliver Cambridge Lower Secondary.

Visit **www.cambridgeinternational.org/lowersecondary** to find out more.

HODDER EDUCATION
e: education@hachette.co.uk
w: hoddereducation.com

ISBN 978-1-398-30130-6

MIX
Paper from responsible sources
FSC™ C104740